Gise

Innovation Management

Giselle Rampersad

Innovation Management

Managing Technology and Networks

VDM Verlag Dr. Müller

Impressum/Imprint (nur für Deutschland/ only for Germany)
Bibliografische Information der Deutschen Nationalbibliothek: Die Deutsche Nationalbibliothek verzeichnet diese Publikation in der Deutschen Nationalbibliografie; detaillierte bibliografische Daten sind im Internet über http://dnb.d-nb.de abrufbar.

Coverbild: www.purestockx.com

Verlag: VDM Verlag Dr. Müller Aktiengesellschaft & Co. KG
Dudweiler Landstr. 99, 66123 Saarbrücken, Deutschland
Telefon +49 681 9100-698, Telefax +49 681 9100-988, Email: info@vdm-verlag.de
Zugl.: Adelaide, University of Adelaide, Diss., 2008

Herstellung in Deutschland:
Schaltungsdienst Lange o.H.G., Berlin
Books on Demand GmbH, Norderstedt
Reha GmbH, Saarbrücken
Amazon Distribution GmbH, Leipzig
ISBN: 978-3-639-21271-6

Imprint (only for USA, GB)
Bibliographic information published by the Deutsche Nationalbibliothek: The Deutsche Nationalbibliothek lists this publication in the Deutsche Nationalbibliografie; detailed bibliographic data are available in the Internet at http://dnb.d-nb.de .

Cover image: www.purestockx.com

Publisher:
VDM Verlag Dr. Müller Aktiengesellschaft & Co. KG
Dudweiler Landstr. 99, 66123 Saarbrücken, Germany
Phone +49 681 9100-698, Fax +49 681 9100-988, Email: info@vdm-publishing.com

Printed in the U.S.A.
Printed in the U.K. by (see last page)
ISBN: 978-3-639-21271-6

Table of Contents

List of Figures

List of Tables

List of Abbreviations

Abbreviation	Term
α	Cronbach's Coefficient Alpha
AGFI	Adjusted Goodness-of-Fit
AMOS	Analysis of Moment Structures
ARA	Actor – Resource – Activity
ARC	Australian Research Council
B/N	Biotechnology/Nanotechnology
CFA	Confirmatory Factor Analysis
CFI	Comparative Fit Index
CRC	Corporative Research Centre
DEST	Department of Education, Science and Training
DNA	Deoxyribonucleic Acid
ICT	Information and Communications Technology
IEEE	Institute of Electrical and Electronic Engineers
IMP	Industrial Marketing and Purchasing
IP	Intellectual Property
NM	Network Management
NPD	New Product Development
R&D	Research and Development
RBV	Resource Based View
RM	Relationship Marketing
RMSEA	Root Mean Square Error of Approximation
RNA	Ribonucleic Acid
SNA	Social Network Analysis
SEM	Structural Equation Modelling
SRMR	Standardized Root Mean-square Residuals
TCE	Transaction Cost Economics
TLI	Tucker Lewis Index
TT	Technology Transfer
UNCTAD	United Nations Conference on Trade and Development

Acknowledgements

I will first like to thank my doctoral supervisors Professor Pascale Quester and Dr Indrit Troshani for their positive and supportive supervision. I am indeed grateful to them for giving me this tremendous opportunity to pursue research in an area that I had a deep interest and passion for quite some time. The combination of Professor Quester's inspiring enthusiasm and energy and Dr Troshani's commendable meticulousness and trustworthiness has been instrumental in the progress of this research. I value their ongoing responsiveness and guidance as these have undeniably helped me in developing my articulation, clarity and overall academic roundedness. It has genuinely been an honour and enjoyable experience to share this PhD journey with them.

I will also like to specially thank my dear friend and mentor, Dr Peter Burns at Adelaide University for his profuse kindness. He went beyond the call on every occasion in assisting me in so many ways on the research and personal levels. He has been a good listener and helped me in developing ideas, understanding areas that I have a keen interest in, and moreover, always generating an intellectual energy that has been essential in stretching me during this journey.

My friends and family who have touched me in so many ways over the years and shaped me also deserve appreciation during this time. I must acknowledge my Durham University Master's supervisor, Dr Elizabeth Burd who generously opened her life and space to me and welcomed and motivated me to pursue the academic life.

I dedicate this book and I will like to give most thanks to my parents, Richard and Gail Rampersad and my sister Ria Rampersad whose unconditional, unshakeable and ever present love has always kept me positive, strong, alive, confident and truly blessed. On that note, I must give thanks to my God as my efforts seem futile without His help and guidance. The PhD process has really been a fulfilling and inspiring one. It has been absolutely essential in preparing me for my future mission.

1.0. Chapter One - Introduction

1.1. Overview

The study is based on the area of innovation networks which is highlighted in Section 1.2. More specifically, the research topic focuses on identifying the key factors for managing these networks and determining the process through which they lead to successful technology transfer (TT). This is discussed in Section 1.3. As Section 1.4 elaborates, the scope of the study focuses on TT and network management (NM). Consequently, the study contributes towards advancing theory development in these two fields and also provides management with useful implications for effectively managing collaborative innovation. These contributions are presented in Section 1.5.

1.2. Background to the Research

The importance of networks has been recognized in strengthening the innovation capacity of a country and in achieving increased international competitiveness (Auster, 1990; Charles and Howells, 1992; Furman et al., 2002; Heikkinen and Tahtinen, 2006; Niosi, 2006; Ruttan, 2001; Tushman, 2004). These innovation networks can be defined as groups of loosely interconnected organizations including universities, research organizations, businesses and government agencies that share scientific discovery and application goals (Dodgson, 1993; Moller and Rajala, 2007). The advent of globalization has intensified the ease of TT across borders and these networks play an instrumental role in enabling countries to nurture their innovation capabilities towards obtaining revenues for their innovations rather than solely paying for foreign innovations (Gans and Hayes, 2004). Many developed countries such as the United States of America, Australia, and the United Kingdom have recognized the importance of networks in building their innovation infrastructure and have, consequently, emphasized the need for multiple organizations across sectors to collaborate in bidding for public R&D funding (Corley et al., 2006; DEST, 2006; DTI, 2007; Plewa, 2005). Similarly, it is crucial for lesser developed countries to strengthen their innovation networks as policies that aim solely towards industrialization and efficiency gains and ignore innovation may contribute to the widening of economic divides (Ruttan, 2001; UNCTAD, 2005):

Most low-income countries do not participate in global research and development networks, and consequently do not reap the benefits that they can generate (Kofi Annan cited in UNCTAD 2005, p v).

Therefore, irrespective of their different development histories, these networks are important in strengthening the capacity of countries to allow them to partake favourably in the innovation-driven global economy (Etzkowitz and Leydesdorff, 1998).

1.3. Research Topic

The process of innovation is both constrained and enabled by the network in which it is embedded (Hakansson, 1987 cited in Ford and Johnsen, 2000). As such, networks can serve as an enabler: they may be synergetic and provide access to new markets, knowledge and resources, and the sharing of risks and costs (Barringer and Harrison, 2000; Wilkinson et al., 2004). However, they can also act as a constraint: relationships can be demanding, sensitive information may be lost and intellectual property may create contention (Ford and Johnsen, 2000). These network inefficiencies have been described as 'black holes' by Hedaa (1999). Therefore, effectively managing networks is important in the innovation and TT process. This study focuses on the research question:

What are the key factors in the management of innovation networks and how do they lead to successful TT?

1.4. Scope of the Study

The three main demarcations of this study in terms of research area, literature and industries are defined in this Section. This helps determine more precisely the contribution of this study.

1.4.1. Demarcation of the Research Area

This study is based on *TT* and not on *diffusion of innovations* and *adoption*. A consensus on the nature and scope of the relationships among these processes is not apparent from the literature. Some authors view diffusion and adoption as components of the TT process, particularly in cases of international TT (Cohen, 2004). Others use these terms interchangeably (Di Benedetto et al., 2003). Diffusion can be defined as the 'spread of a new idea from its source of invention or creation to its ultimate users or

10

adopters' and adoption is defined as 'the decision to continue full use of an innovation' (Rogers, 1962, p 19). Although some authors view diffusion as an inter-organizational process (Muzzi and Kautz, 2004), in this study, a clear distinction is made between them: TT is an inter-organizational process whereas diffusion and adoption generally focuses on end-users. This distinction is made because this study focuses on the dynamics and effectiveness of inter-organizational networks rather than on end-user networks. Therefore, this study will not utilize adoption and commercialization models that pertain to end markets as in Moore (2002) and Jolly (1997).

1.4.2. Demarcation of the Literatures

This study focuses on the TT and inter-organizational literatures particularly the Industrial Marketing and Purchasing (IMP) literature. That said, the TT stream is positioned within the broader literature on innovation and technology management, and therefore, they will also be incorporated when relevant. Similarly, although this study draws heavily on the IMP literature as it has placed pronounced importance on NM, it also draws on the wider inter-organizational literature as the IMP literature is still evolving and is being influenced by a number of related fields. The inter-organizational theoretical literature is multidisciplinary and spans from the economics to the behavioural fields (Barringer and Harrison, 2000).

Given the novelty of the study and the elementary stage of theory development on NM in the IMP literature, this study taps into a number of these relevant streams in particular social network analysis, relationship marketing and triple helix. These fields were chosen given their relevance to the network perspective, and hence, theories that focus on an organizational perspective such as the resource based view have not be used (Barringer and Harrison, 2000). Furthermore, as a network perspective is by definition interactive, dynamic, highly complex, it requires a more systems thinking approach whereas linear theories, such as the value chain model were found to be unsuitable for this study (Etzkowitz and Leydesdorff, 1998; Powell et al., 1996).

1.4.3. Demarcation of the Industries

This study focuses on innovation networks from four industries in the first instance. These include biotechnology, ICT/defence, automotive and wine and were chosen because of their national importance to Australia as well as their linkages and relevance regionally and internationally. The biotechnology industry (specifically biotechnology/nanotechnology) was investigated because of its potential for Australia which has been historically strong in life sciences research. Biotechnology has

11

been highlighted in Australia's national research priorities under the development of frontier technologies (ARC, 2006; DEST, 2006). ICT and defence are also key industries in achieving these priorities. The defence-related ICT industry was examined in this study because South Australia has the highest concentration of ICT professionals in the southern hemisphere with over 9000 people working in the defence industry, making it an ideal innovation cluster. The ICT industry is important globally as a recognised enabler of development and economic growth. The automotive industry is another industry that was investigated because of its potential to strengthen regional synergies. Strengthening the supply of automobile components to the well-positioned automobile industry in Asia, is especially important as the sale of cars in China increases (Austrade, 2007). The wine industry was also be researched as it has been successful internationally with linkages in the United States, New Zealand, France and South Africa and is a large source of exports (Austrade, 2007).

This study may help in improving the effectiveness of collaborative innovation in these industries. After the preliminary, exploratory interviews, three of the four industries were selected based on the size of their networks for further quantitative work. One of the three networks was used for a pilot study and the remaining two were used to provide the context for the field work. If patterns in the results emerge in dissimilar industries during the latter stage, it may be possible to generalize the results beyond those under investigation to related industries. Differences may provide industry specific implications.

1.5. Significance / Contribution of the Research

Although the literature on inter-organizational networks is extensive, it focuses predominantly on the advantages of such networks, ignoring their disadvantages (Hedaa, 1999). Ironically, this over-optimism may also be reflected in the sharp increase in inter-organizational innovation networks, despite their high failure rates of over 50% (Barringer and Harrison, 2000; Ford and Johnsen, 2000; Ojasalo, 2004; Park and Ungson, 1997; Porter, 1987; Wilkinson et al., 2004). Consequently, further research is needed on managing innovation networks, especially to improve their effectiveness so that their anticipated advantages could be realized.

To date, however, no articulated theory about NM that addresses these issues has been found (Blankenburg et al., 1999; Dyer and Nobeoka, 2000; Ford et al., 2002; Gulati, 1999; Jones et al., 1997;

Moller et al., 2002; Ramirez, 1999). The IMP literature includes few approaches to NM which are generally conceptual or descriptive in nature, and yet, to be tested empirically (Ojasalo, 2004). Additionally, in terms of methodology and level of analysis, empirical network studies have focused on focal organizations and dyads rather than on adopting an overall network perspective which is not limited to that of a particular network actor (i.e. an actor is an organization in the network). Existing studies are limited in addressing the need for coordinating a group of actors and the complexities for achieving network efficiencies (Moller et al., 2002). This study aims to provide empirical evidence to advance NM theory by adopting a network perspective rather than that of any of its focal organizations or dyads. In doing so, it develops valid scales appropriate for the network level of analysis that may be applied in future studies to advance the field.

Similar to the NM literature, the existing TT literature adopts the view of a focal organization or relationship between the transferor and recipient. The management of TT could be aided by analysis from multiple perspectives – individual, organizational, relationship and network as shown in Figure 1. Although the extant TT literature focuses on the relationship or organizational perspective, this study provides a network perspective of TT. The latter perspective is relevant because in many cases, more than two actors are involved in innovation and TT initiatives (Etzkowitz and Leydesdorff, 1998).

Figure 1. Multiple levels of analysis of TT

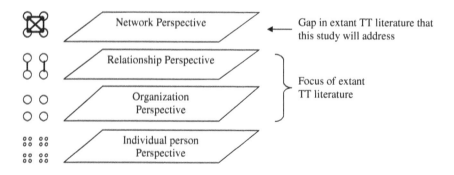

13

Interaction and communication at the network level may affect the effectiveness of TT, depending on the characteristics of the technology to be transferred (Lin and Berg, 2001). Therefore, this study will apply a network perspective to the TT context in order to provide management with recommendations on how to effectively manage TT in an environment that is increasing in complexity due to globalization. It will, therefore, advance theory development in enriching TT knowledge with NM principles.

Useful management implications are derived from this study for actors involved in innovation networks such as policy makers, businesses and universities. The study explores the importance of network factors such as power distribution, effective coordination of the network and harmony in the TT process and empirically tests their impact on network efficiencies and overall effectiveness.

1.6. Summary

This study investigates success factors for managing networks involving TT. Its scope is focused on an inter-organizational network level perspective. It also initially examines four industries – wine, automotive, biotechnology / nanotechnology and ICT. Further quantitative work is then conducted in the latter two industries. Findings from this research may be of interest to a variety of players involved in these networks including government agencies, businesses and universities. Additionally, this multidisciplinary research contributes to theory development in both fields of NM and TT.

2.0. Chapter Two - Technology Transfer

2.1. Overview

The importance of innovation networks to facilitate TT is growing. Although inter-organizational collaboration has long existed, it is increasing in complexity and as such, the management of the underlying networks is important. This chapter first discusses the evolution of TT in Section 2.2. It then defines TT in Section 2.3 and it analyzes its success factors and respective outcomes in Section 2.4. Extant literature investigates success factors and outcomes predominantly from a focal organization or relationship perspective. Although the network perspective is rarely explored empirically, its significance in enriching this research stream has been recognized.

2.2. The Evolution of TT

2.2.1. Antecedents to the Evolution of TT

TT involving universities, government and industry is not new. In the United States, Swann (1988) traces TT among institutions back towards the end of the nineteenth century when universities transferred research to industry, including the US Military Academy, the American Literary, Scientific and Military Academy and Rensselaer Polytechnic Institute and later the Sheffield Scientific School at Yale and Massachusetts Institute of Technology. In comparison to the United States, he argues that university-industry links in Germany were more sophisticated with the former only reaching a similar level of advancement between the two world wars (Swann, 1988).

Although inter-institutional TT has existed for a long time, there has been a recent increase in its rate and in the complexity of innovation networks supporting the process. Charles et al. (1992) provide a compelling argument on the factors that may have contributed to this trend. These included the world wars, which catalyzed industry research in the United States, the United Kingdom and Japan in naval, pharmaceutical, radar and energy-related research respectively. In addition, they identify the recent emergence of enabling technologies such as information and communication technologies (ICT), biotechnology and nanotechnology that have reduced barriers between research institutions and industries as these technologies contribute to a broad variety of industries. Additionally, networks have been recognized for their influence on learning and developing capability (Bessant et al., 2003). Furthermore, the increased complexity of R&D, developmental time and costs; decreased product life cycles; limited availability of scientific expertise and shifts in public R&D funding towards multi-institutional research have lead to increases in R&D networks (Heikkinen and Tahtinen, 2006; Tushman, 2004). By the mid 1970s there was also a shift from predominantly simple bilateral arrangements to more complex research networks (Charles and Howells, 1992).

2.2.2. Theoretical Bases for the Evolution of TT

By the 1980s, the TT literature also reflected the shift from single relationships to more complex research networks. The TT literature is part of the broader literature on innovation, R&D and technology management (Gibson et al., 1990). A recent review by Tushman (2004) of the past 50 years of publications in one of the top journals in this discipline, IEEE Transactions on Engineering Management, also reiterates this trend thereby recognizing the value of inter-organizational networks in R&D research. He argues that the literature was previously disintegrated and focused on a narrow range of topics including project management, creativity, technology strategy and policy, engineering careers and organizational design. Furthermore, he points to the fact that these were inadequate in understanding practical R&D challenges facing firms that a network perspective would explain. Similarly, a bibliographic review of the TT literature from 1985-1990 revealed that 90% of it focused on 'international transfer, lesser developed countries, regulations, ethics, economics and finance' (Gibson et al., 1990, p 279). More recently, over the last 30 years, the literature has included areas such as university-industry relationships (Niosi, 2006). There are currently more studies recognizing the value of inter-organizational networks and the present literature includes terms such as 'triple helix' to describe networks of governments, universities and industry (Auster, 1990; Charles and Howells, 1992; Chesborough, 2003; Heikkinen and Tahtinen, 2006; Laursen and Salter, 2006; Niosi, 2006; Tushman, 2004).

In addition to the shift in the literature towards recognizing the value of inter-organizational networks, there has also been a shift from assessing the importance of TT towards analyzing factors contributing to its effectiveness. Niosi (2006) argues that previously there has been a focus towards studying the impacts of the 1980 Bayh-Dole Act in the United States, and the value of university-industry research via mechanisms such as spin-offs, clusters and TT offices. He calls for future research that moves beyond this towards increasing the effectiveness of the TT process.

16

2.2.3. Approach to Scope

Although the importance of a network perspective in improving the TT process has been recognized, such literature has been predominantly descriptive or conceptual. Previous empirical research on TT focuses on either unilateral or at most bilateral types of collaboration (Gallagher, 2004; Medlin, 2001; Plewa, 2005; Rebentisch and Ferretti, 1995). Therefore, given the trends towards increasing complexity of TT, the scope of this study is on a network perspective as it offers a more comprehensive and realistic view of the process.

2.3. Definition of TT

TT generally addresses the inter-organizational movement of knowledge. This common theme underlines various specialized definitions that have been adopted at various sectors and disciplines. TT is defined as 'the process of moving innovations from their origin to their point of operation' (Plewa, 2005, p 60) and 'the movement of any type of scientific or technological knowledge from one sector to another' (Ryan, 2004, p 35). It is sometimes viewed in its intra-organizational context. This study views TT as an inter-organizational process between supply and recipient organizations (Kedia, 1988). Therefore, integrating the aforementioned definitions, TT is defined in this study as *the movement of innovative technologies from the transferor to the recipient organizations.*

Transfer Scope

The characteristics of the technologies to be transferred vary. Williams et al. (1990) define *technology* as capability or 'knowledge embodied in an artefact (software, hardware or methodology) that aids in the accomplishing of some task' (Williams and Gibson, 1990, p 45). They deliberately exclude knowledge stored in a person's memory and only focus on knowledge in a communicable form as in product or process innovations. The wine industry illustrates these types of innovations. An example of a product innovation can be the spectronics device recently developed by the University of Adelaide to test the colour of red wine and to offer a significantly more affordable testing product compared to its alternatives (CRC-V, 2006). An example of process innovation involves the improvement in the irrigation process, from flood irrigation to more efficient and cost effective drip techniques.

The technology to be transferred can also be categorized based on the degree of R&D required. Innovation may be viewed as ranging from radical to incremental with the former requiring higher R&D investments and having greater proprietary potential (Ojasalo, 2004). Radical or breakthrough innovations involve the 'first generation of an entirely new product or process' whereas incremental or

17

derivative innovations 'refine or improve selected performance dimensions to better meet the needs of specific market segments' (Clark and Wheelwright, 1993, p 105).

2.4. Success Factors of TT

Several factors contribute to successful TT. The analysis of these factors could be undertaken at different levels: they could be investigated at the level of the individual, organization, relationship and network. Extant TT literature provides an analysis predominantly from the perspective of a focal organization or relationship (see Table 1). Exploring a network perspective may be useful in extending the view of management beyond its immediate view to indirect dynamics that may affect TT success.

Table 1. Success Factors of TT

Perspective	Success factors	Authors	Methodology	Future work
Individual person	National culture	(Garrett et al., 2006; Hofstede, 1980; Kedia et al., 2002; Kedia, 1988; Lin and Berg, 2001; Song and Thieme, 2006; Wiley et al., 2006)	There have been preliminary empirical studies on the impact of culture on TT. However, they have provided analysis on the relationship or organization level. These have produced conflicting results.	An analysis of the impact of culture on TT is best done at the individual level. Factors other than culture influence human behaviour. A network perspective is not seen as appropriate for exploration.
Focal Organization	Adaptive ability, knowledge architecture	(Gallagher, 2004; Rebentisch and Ferretti, 1995)	Some articles have been conceptual. However, others have provided empirical evidence. The literature has established that certain organizational and relationship factors impact on TT.	The effect of informal sub-networks between organizations in initiating innovation projects and championing innovation has not been explored.
Relationship	Organization culture, motivation	(Gallagher, 2004; Kedia, 1988; Plewa, 2005; Medlin, 2001)		
Network	Network level communication and its interaction with transfer scope	(Auster, 1990; Charles and Howells, 1992; Heikkinen and Tahtinen, 2006; Lin and Berg, 2001; Niosi, 2006; Rebentisch and Ferretti, 1995; Tushman, 2004)	Although the network perspective has been preliminarily applied to TT, no empirical evidence has been found on the management of TT from such a viewpoint.	Exploring TT success factors from a network perspective would enhance the research stream. Such factors may include coordination, harmony and network efficiencies such as communication and R&D.

2.4.1. Individual Perspective

National Culture Difference

Characteristics of individual persons may have an impact on TT such as national culture. Studies about the impact of national culture on TT outcomes have led to conflicting results. On the one hand, studies done by Lin et al. (2001), Garrett et al. (2006) and Song et al. (2006) provide empirical evidence supporting the impact of national culture on TT. Lin et al. (2001) argue that national culture difference is a mediating factor between the nature of technology and TT outcomes as it may contribute to communication difficulty between technology transferor and recipient. However, their study adopted a simplified approach representing culture as either 1 or 0 and did not engage with the multidimensional complexities of culture as provided by Hofstede (1980), such as, uncertainty avoidance, power distance, degree of individualism and collectivism, orientation towards masculinity and femininity, and abstractive and associative. Similarly, Low and Chapman (2003) recognized the convergence of the literature on dimensions of culture of temporal, relationship, confidence and communication. Garnett el al.'s (2006) study incorporated some dimensions. However, like Song et al. (2006), their study focused on innovation internal to an organization rather than inter-organizational networks. Unlike in the aforementioned studies, Wiley et al. (2006) provided empirical evidence that indicates that national culture does not impact on inter-organizational relations. Nevertheless, all four studies adopted a relationship or organization perspective rather than a network perspective.

The investigation of the impact of national culture on TT outcomes may not be appropriate at a network level as it is an individual characteristic. In an increasingly globalized world, networks may comprise firms from multiple countries and also firms with employees of different nationalities. Furthermore, factors other than national culture may determine an individual's behaviour, such as, gender, education, experience, age, organization culture and leadership, stage of economic development in a country and its related technological capacity (Garrett et al., 2006; Medlin, 2001). 'National culture explains between 25% and 50% of the variation in an individual's behaviour' (Gannon, 1994 cited in Garrett et al., 2006, p 305). Therefore, although extant literature does not provide an analysis of the impact of national culture from a network level per se, this may not be appropriate as it may lead to inconclusive results, given that national culture is only a partial determinant of individual behaviour.

19

2.4.2. Focal Organization Perspective

Differences in Knowledge Architecture

The impact of knowledge architecture on TT outcomes has been established by adopting an organizational perspective. A knowledge architecture is 'a characterization of the structure and the artefacts into which knowledge has been embodied in the organization, and describes the way an organization stores and processes information' (Rebentisch and Ferretti, 1995, p 10). It includes technologies, operating procedures, social and organizational relationships and organizational structure. These are important in transferring technologies (Kedia, 1988; Mathews, 2001).

Organizational Adaptive Ability / Absorptive Capacity

The impact of an organizational absorptive capacity or adaptive ability on TT outcomes has been established by adopting the perspective of a focal organization capacity (Baranson, 1970, Driscolli and Wallender, 1981, Dunning, 1981 cited Kedia, 1988; Mathews, 2001). An organizational adaptive ability is the capacity of the recipient organization to integrate transferred technologies (Cohen and Levinthal, 1990; Gallagher, 2004; Rebentisch and Ferretti, 1995).

2.4.3. Relationship Perspective

Organizational Culture and Motivation Difference

The effect of organizational culture and motivation on TT outcomes has been established at the relationship level of analysis. An organizational culture is distinctive to one organization, organizational unit or group and may comprise flexibility, time orientation, market orientation, empowerment, organizational compatibility and experience (Medlin, 2001; Plewa, 2005). Motivation includes collaborative goals and future orientation on outcomes (Medlin, 2001).

2.4.4. Network Perspective

Very little has been done in relation to networks, and only a few concepts or variables have been identified in the past literature, including coordination, harmony, and power distribution (Charles and Howells, 1992), as well as communication (Lin and Berg, 2001; Williams and Gibson, 1990).

Transfer Scope

Other researchers have argued, albeit not in a network context, that the process by which technology transfers is moderated by other factors, including transfer scope (Lin and Berg, 2001; Rebentisch and Ferretti, 1995). However, none of these studies has provided empirical evidence and these stated relationships remain, at best, conjectural. Section 2.3 distinguished between product and process innovations. Kedia et al. (1988) argue that product-embodied technology is relatively easy to transfer as it consists of the product itself compared to process-embodied technology that consists of more tacit knowledge compared to explicit knowledge (Kedia et al., 2002; Kedia, 1988). It should be noted that the literature on the diffusion of innovation and adoption investigates this occurrence. It has been established that there is an impact of the interplay between social systems and technological characteristics including relative advantage, compatibility, complexity, divisibility and communicability on diffusion and adoption (Rogers, 1962). Nevertheless, as demarcated in Section 1.4.1, this study, like the majority of studies on TT, focuses on inter-organizational TT, whereas diffusion of innovations and adoption focus on end-users. Therefore, further work is required on this interplay between technological factors and network level processes in the TT field.

2.5. TT Outcomes

Measurement of outcomes is important in moving discussion of theoretical assumptions closer to practice (Adams et al., 2006; Chapman and Magnusson, 2006; Soosay and Chapman, 2006). A measure for evaluating the success of TT that reflects the diversity of perspectives and goals of the different actors involved in the TT process remains absent from the literature. Szakonyi (1994) evaluates thirty years of research on R&D management and argues that given this variety of perspectives, there is no consensus on measures used to evaluate outcomes. Spann et al. (1995, p 19) also argue that 'the measures or metrics of technology transfer effectiveness are neither well defined nor universally accepted'. They maintain that TT has various outcomes, stages, scopes, durations and that various actors may value different outcomes given their roles at different stages in the TT process. Additionally, as TT can be analyzed on multiple dimensions, outcomes can be assessed from the perspective of individuals, organizations, relationships and networks.

Existing measures of TT success are mainly limited to a focal organizational perspective with a specific role in the TT process. These do not capture the complexity of perspectives and roles of actors operating in a network context of TT. They vary in the level of focus on behavioural, technical, economic, quantitative and qualitative aspects. Many studies adopt the perspective of the recipient organization in the TT process. An example of a qualitative measure with a behavioural focus is given in DiBeniditto et al.'s (2003) study where the Technology Acceptance Model (TAM) was applied and behavioural intention to adopt was identified as a suitable measure. However, adoption rates are most applicable as a measure to the recipient organization rather than encompassing the objectives of the multiplicity of players involved in TT. Measures of technical effectiveness are also used to assess TT success from an organizational perspective. These include technical effectiveness of transferee compared to the transferor; effectiveness compared to the transferee's other projects; technical effectiveness compared to plan and technical effectiveness compared to major competitors (Lin and Berg, 2001). However, these measures are very rigid and the comparison criteria may not be applicable in all cases e.g. some transferees may not have other projects. Measuring milestone achievement is common (Spann et al., 1995) but this fails to capture diverse qualitative and quantitative goals involved in TT. Milestone achievement is also associated with project deliverables rather than the more continuous nature networks at the core of this study. Garcia-Valderrama et al. (2005) use Tipping et al.'s (1995) technology pyramid that comprises both qualitative and quantitative measures such as financial ratios. However, these measures continue to predominantly analyze TT success from a focal organization perspective.

Other studies provide an analysis of TT effectiveness from the perspectives of several organizations that possess limited number of specific roles. These fail to capture the complexity of perspectives and roles of actors operating in the network context of TT. Spann et al. (1995) provide a preliminary analysis from the perspective of sponsors, developers and adopters. However, the definition of these roles is not clear-cut as the same player may adopt various roles within a network. Also, there are other roles (e.g. government policy makers or university administrators) in the TT process that have not been explored. The definition of roles and their desirable measures are debatable and incomplete.

In addition to measures of TT effectiveness from focal organization perspectives, other extant measures of R&D effectiveness are also inadequate in a network context as they focus on the departmental level of analysis. Garcia-Valderrama et al. (2005) propose a 35-measure construct of R&D effectiveness that assesses R&D inputs, processes, outputs and results. However, it is very specific to the performance of the R&D department in contributing to the achievement of an organization's objectives.

22

R&D Efficiency

Any measure evaluating TT success in a network context should cater for the variety of outputs that the different actors attain, given the levels of inputs that they contribute to the collaboration. R&D efficiency provides a relative measure of R&D outputs compared to R&D inputs (Fritsch, 2004; Fritsch and Meschede, 2001; Fritsch, 2000). Unlike R&D expenditure, it offers a relative measure of inputs to outputs, and thus, facilitates comparisons between TT initiatives. In the network context, it is relevant as actors contribute inputs to the TT process including funding, infrastructure, skills or other resources and seek to obtain outputs through the TT process.

2.6. Summary

This study examines TT from a network perspective. In contrast, the extant TT literature adopts a focal organizational or relationship perspective and explores factors such as knowledge architecture, adaptive ability, organizational culture, motivation and national culture. However, as shown in Figure 2, this study will focus on factors that are applicable from a network level of analysis, such as, the impact of the interplay between transfer scope and network level communication and interaction on R&D efficiency and overall network effectiveness. To assist in the investigation of the network level of analysis, Chapter 3 provides a review of the network management literature.

Figure 2. Conceptual Framework for TT in Networks

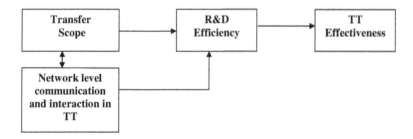

3.0. Chapter Three - Network Management

3.1. Overview

In order to provide a network perspective of TT, the network literature and in particular the relevant IMP literature must be explored. The importance of combining inter-organizational network studies with innovation and TT has increasingly been recognized. Hakansson (1987) appropriately combined the network approach with innovation as he recognized that innovations are generated by interactions amongst network actors (Hakansson, 1987; Hakansson, 1989). Indeed, in 1991, a special issue of the Research Policy journal was devoted to this topic (Iacobucci, 1996). Subsequently, the International Journal of Technology Management called for special papers on innovation networks in 2006 and the term 'network of innovators' was coined (DeBresson and Amesse, 2006). This chapter discusses the evolution of NM by first identifying some of its diverse antecedents. It then highlights theoretical bases of NM and the researchers' incapacity to establish causality between success factors and outcomes as well as the inadequate analyses often undertaken at the network level.

The network field is being shaped by a major debate on whether networks can, in fact, be managed as Section 3.2 elaborates. At the crux of this debate are two main questions of (1) the level of analysis that the researcher adopts as Section 3.2.3 addresses, and (2) the ontological characteristics of a network, as Section 3.3 explains. Application of the NM literature to a TT context is not straightforward as the IMP literature remains elementary in its NM theory development, although NM is fast evolving (Golfetto et al., 2007; Moller and Rajala, 2007). Nevertheless, this chapter identifies success factors in NM that are prominent in the literature in Section 3.4, as they may be incorporated to advance both NM and TT theory development.

3.2. The Evolution of NM

3.2.1. Antecedents to the Evolution of NM

The network literature is extensive and fragmented and can be found in many social sciences fields including finance, economics, geography, international business, entrepreneurship studies, strategic management, marketing, political science, and sociology (Barringer and Harrison, 2000; Iacobucci, 1996). Although there is much overlap in these respective theories, a unified view of the phenomenon has not yet been provided, and hence, the theories should not be interpreted as either equivalents or alternatives, but as providing different perspectives (Axelsson and Easton, 1992). Although the majority of network research can be traced to social anthropology, a branch of structural sociology, it has been developed using varying methodological approaches (Araujo and Easton, 1996).

In general, there has been a shift in methodological approaches employed in network research from structuralist to process orientation, which is in favour with the recognition of the need for NM research that demands an understanding of processes. Galaskiewicz (1996) has highlighted this shift from formal mathematical analysis of structure to model building and hypothesis testing. He argues that in the mid-1970s, there was a focus on using mathematical models and graphs to describe the structure of networks (Allen, 1987; Sonquist and Koenig, 1975; White et al., 1976). Researchers then started to link network structural characteristics, such as, centrality and density, to network outcomes (Boje and Whetten, 1981; Cook, 1977; Knoke, 1983). Although quantitative methods were used, they mainly served to describe actors' positions and network structure rather than to explain network processes (Salancik, 1995). Tables 2 and 3 illustrate the structural dimensions of networks that have been explored.

Table 2. Structural Dimensions of Networks - Network as Focus of Analysis adopted from Auster (1990)

Size	Number of organizations in the network
Density	Number of linkages in the network
Diversity	Linkage: number of different types of linkages in the network
	Organizational: number of different types of organizations in the network
Reachability	The number of links separating two organizations
Stability	Linkage: whether the form of linkage in the network remains the same over time
	Organizational: whether the organizations in the network remains the same over time
	Frequency of change: how often linkages or organizations change
	Magnitude of change: how many linkages or organizations change
Stars	The number of organizations with greater then X number of ties
Isolates	The number of organizations with no linkages to other organizations
Linking pins	Organizations with extensive and overlapping ties to different parts of a network

Table 3. Organizational Position within a Network as Focus of Analysis adopted from Auster (1990)

Centrality	The proportion of the sum of relations that involve organization X
Range	The number of contacts organization X has
Multiplexity	The extent that organization X is connected to a high proportion of organizations in the network by multiple types of relations
Degree of Horizontal Interdependence	The number of linkages with organization X at the same stage of the transformation process
Degree of Vertical Interdependence	The number of linkages with organization X at different stages of the transformation process
	Backward: the number of linkages with organization X at an earlier stage of the transformation process
	Forward: the number of linkages with organization X at the later stage of the transformation process

While a focus on structure may be useful to some extent, it does not address process factors, which are fundamental for managing such networks. By the late 1970s, researchers started to use network analysis as a secondary method to others, such as, resource dependency, institutional theory, transaction cost economics and social exchange which were seen as more useful in explaining processes (Galaskiewicz, 1996).

Unlike its counterparts in many other disciplines, the industrial networks literature has a process orientation. Some bodies of literature found in sociology, economic geography and international business, such as, social networks and comparative studies have a strong focus on structure, including the use of socio-metric techniques (Araujo and Easton, 1996). The IMP group was formed in 1976 by researchers from 5 European countries (Gadde and Hakansson, 2001). The focus of IMP work is generally around industrial networks, which in comparison to social or electronic networks include 'actors involved in an economic process which convert resources to finished goods and services for consumption by end users' (Axelsson and Easton, 1992, p xiv). The orientation of IMP is inclusive, and although its literature focuses on process studies, it contains investigations of structure (Laine and

Kock, 2000; Salmi et al., 2000). It is evolving and incorporates theories and researchers from different disciplines.

As IMP researchers share different disciplinary backgrounds, their work has been cross-referenced with a variety of disciplines. The IMP literature emerged from theories outside the marketing field such as inter-organizational, social exchange and new institutionalist theories as well as earlier and emergent trends in the marketing and purchasing areas (Araujo and Easton, 1996; Hakansson, 1982). Based on the influence from the inter-organizational literature, the IMP literature views the organization as dependent on others for resources, and as such, the characteristics of organizations have been investigated as they relate to each other and the links and interaction among them in terms of their formalization, intensity and standardization (Hakansson, 1982).

The major influence from the social exchange theory was that of connectedness (Araujo and Easton, 1996). In terms of the influence from the New Institutionalists, a line of thought within the micro-economics literature including transaction cost economics (TCE), the IMP literature recognizes that transactions could take place internally within an organization or within a market (Williamson, 1975). TCE focuses mainly on economic factors (Auster, 1990; Barringer and Harrison, 2000) and, thus, the IMP literature has broadened beyond this view towards social aspects. Trends from the marketing and purchasing fields that have influenced the IMP literature reflect the thinking that both buyers and sellers are viewed as market participants engaged in a long term relationship in the context of continuous raw material supply, and that become institutionalized into set roles based on the other party's expectations (Hakansson, 1982).

Due to rich linkages with other disciplines and its orientation that includes process investigation in an inter-organizational context compared to its counterparts (Araujo and Easton, 1996), the IMP approach provides a suitable basis upon which NM ideas have naturally begun to be explored. As the IMP literature has cross-references with other theories, it can build on concepts introduced in these disciplines in order to develop its theories. Additionally, network studies have been moving away from personal networks to inter-organizational networks, and as such, these studies have been pushed towards industrial network approaches which focus on innovation in the inter-organizational context (Iacobucci, 1996). The importance of advancing NM has also been stressed in the IMP literature (Campbell and Wilson, 1996; Ford and Johnsen, 2001; Ojasalo, 2004).

27

3.2.2. Theoretical Bases for the Evolution of NM

Although models of NM are present in the IMP literature, theory development remains embryonic, although this is fast changing. The IMP literature is evolving as its focus switches from the dyadic level to the network level of analysis.

The first major contribution of the IMP group has been the interaction approach, focusing on the inter-organizational relationship at the dyadic level. This has been followed by the industrial networks approach that has moved beyond the dyadic to the network level of analysis as it emphasizes the importance of long term, stable relationships in industrial markets (Axelsson and Easton, 1992). The Actor-Resource-Activity (ARA) model then emerged which touched on the relationships between network characteristics of actors, resources and activity structures and network efficiency (Gadde and Hakansson, 2001). Although it provides a conceptual understanding of these relationships, it does not provide empirical evidence of the relationships between network characteristics, processes and efficiencies. Currently, there is increasing attention being placed on this.

Although no comprehensive NM theory currently exists, some conceptual models have emerged. These are being shaped by the major debate on whether networks can be managed. Arguments in this debate differ based on the researchers' views towards the ontological characteristics of networks and the levels of analysis adopted. Traditional researchers from the industrial network approach tend to view networks as boundary-less phenomena (Hakansson and Ford, 2002; Hakansson and Snehota, 1995) that cannot be managed. The NM model by Ford et al. (2002) that emerged from this approach adopts the view that although it may be impossible to manage networks, 'managing in' networks may be possible by coping, reacting and managing relationships. This model adopts the perspective of a focal organization operating in a network and offers advice on formulating network pictures, adopting strategies and recognizing various levels of outcomes to the organization, its relationships and the network in which it is embedded. However, the NM model does not establish causality between components (Ford et al., 2002). Establishing causality through empirical testing may increase the validity of a model and advance theory development.

Within the traditional industrial network approach, role theorists have also attempted to apply the concept of roles to understand network dynamics (Anderson et al., 1998; Havila, 1992; Knight and Harland, 2005). Arising from this approach, Heikkinen et al. (2007) highlight several roles for managing in nets. While their study does offer insightful suggestions on management roles in nets, similar to its aforementioned counterparts, it does not relate the roles with net outcomes, and thus, it does not provide an indication on which roles are most significant managerially and theoretically (Heikkinen et al., 2007).

Contrary to the traditional industrial network theorists and the subset of role theorists, other scholars from the strategic/value networks approach attempt to identify specific sub-networks defined around strategic issues termed as issue-based nets or value nets which, they argue, could be managed (Brandenburger and Nalebuff, 1996; Brito, 1999; Jarillo, 1993; Moller and Rajala, 2007; Parolini, 1999). They focus on analysis at the net level rather than an organizational perspective. Models emerging from this approach classify nets based on their value proposition and suggest management strategies for each type of network (Moller and Rajala, 2007; Moller et al., 2002). This analysis does have limitations, as networks are dynamic and difficult to classify. Additionally, it is often based on scant empirical testing (Moller and Rajala, 2007; Moller et al., 2002) and strategies recommended have yet to be conclusively linked to network outcomes.

3.2.3. Approach to Scope

Although the IMP literature provides rich conceptual multi-layered analysis based on different levels of aggregation of units within the network, the network level perspective remains underdeveloped empirically. The literature generally adopts an organizational perspective based on network involvement with little attention given to the whole network (Provan and Milward, 1995). Measures, constructs and operational definitions given in the literature remain biased towards organizational antecedents and outcomes rather than reflecting sufficient network level measurement. However, growing importance has been and is being given to assessing whole networks and their effectiveness in both academic and government policy-making quarters (Jensen et al., 2007). Increasingly, recognition has been made that certain public services and national priorities, such as, innovation goals could only be accomplished through the cooperation of multiple organizations. In the US for instance, the past three decades could be credibly termed the 'era of inter-institutional research collaboration' (Corley et al., 2006, p 975). US technology policy has shifted from supporting small research projects to inter-organizational, block grant research. Similarly, many other countries, such as the UK, Canada and Australia, have incorporated the need for multiple organizations to cooperate in bidding for research grants. Therefore, a focus on the network level as a whole is important to ensure that strategic goals are achieved. This is especially significant to ensure that scarce public funds are spent efficiently (Provan and Milward, 1995). As such, the scope of this study will be on NM from the point of view of the whole network rather than that of a focal organization.

3.3. Definition of networks

Networks are broadly defined as 'a set of actors and the relational ties between them' (Iacobucci, 1996, p 392). However, it is rightly argued that networks can be studied without focusing on relational ties, although they are present. Networks have the advantage of allowing researchers to choose the level of aggregation they wish to adopt within the network based on the research questions.

Networks are studied as they reflect the complex realities that organizations face:

> Networks provide rich and complex metaphor for economic exchange relationships. They offer an opportunity for managers to understand the complex set of relationships which managing a modern business organization involves. (Axelsson and Easton, 1992)

At the crux of the major debate that is shaping the field is the question, 'what is a network' and more specifically 'what are the boundaries of a network'? Some researchers of the IMP group hold the opinion that networks are boundaryless (Ford et al., 2002; Hakansson and Ford, 2002):

> There is no single, objective network. There is no 'correct' or complete description of it. It is not the company's network. No company owns it. No company manages it, although all try to manage in it. No company is the hub of the network. It has no 'centre', although many companies believe that they are at the centre. (Ford et al., 2002, p 4)

Other researchers argue that sub-networks with definite boundaries can, in fact, be defined (Gulati et al., 2000; Moller et al., 2002; Parolini, 1999). Despite the challenges in defining network boundaries, many organizations have successfully collaborated in networks to achieve their research and development (R&D) and innovation objectives. From the management viewpoint, it is necessary to focus effort on certain meaningful parts of the network (Ojasalo, 2004). These subsets, 'nets' of the larger network also termed strategic nets or value nets have been defined around interrelated groups of actors pursuing shared innovation or TT goals (Heikkinen et al., 2007; Ritter and Gemunden, 2003).

In defining networks, there is confusion on the extent to which networks are intentional or emergent as some argue that this may affect the degree to which NM is possible. Achrol et al. (1999) distinguish between 'network of organizations' and the 'network organization' approaches. Researchers from the former perspective view networks as emergent, and therefore, somewhat unmanageable (Ritter et al., 2004; Stacey, 1996), whereas researchers in the latter perceive nets as intentional, and hence, manageable (Heikkinen et al., 2007; Parolini, 1999). However, such a distinction is not clear cut. Moller and Rajala (2007) from the latter group of researchers argue that even emergent networks arise due to intentional actions of its participants. Similarly, intentionally created networks, such as, interventions involving national systems of innovation may be outlived by continuous networks that emerge in the process (Etzkowitz and Leydesdorff, 2000).

Rather than attempting to distinguish between somewhat ambiguous categories of emergent and intentional networks, focusing on continuous networks may offer useful insights in exploring NM. This approach resonates with relationship marketing scholars whose attention is placed on ongoing, lasting relationships rather than on mere discrete shorter term transactions (Dwyer et al., 1987; Morgan and Hunt, 1994). Similarly, networks can be viewed as 'endless transitions' of continuous interaction between organizations (Etzkowitz and Leydesdorff, 2000; Medlin, 2006) regardless of intentional interventions and emergent periods. Therefore, this study deals with 'live' nets, that is sets of organizations that are actively interrelated, and therefore, really operating together continuously as opposed to being necessarily embedded in a formal, temporal structure that may or may not serve its intended purpose. Hereafter nets and networks would be used interchangeably.

In defining networks, attempts are also made to establish the level of formality among ties in the network and classify them, as some argue that different types of networks may require varying management solutions. Moller and Rajala (2007) categorize networks based on their value proposition. They further define innovation networks as 'relatively loose science and technology-based research networks involving universities, research institutions, and research organizations of major corporations...guided by the ethos of scientific discovery' (Moller and Rajala, 2007, p 900). Although they highlight the relatively loosely coupled nature of these networks, they acknowledge that networks are dynamic and that in reality networks may comprise various forms that change over time. Furthermore, in some countries, governments may adopt heavy-handed approaches to innovation policy and introduce very formal interventions requiring strong ties between organizations (Mani,

31

2002). Nevertheless, innovation networks are also seen by other authors as relatively loosely coupled organizations although there may be strong and weak ties amongst them (Freeman, 1991).

For this study, therefore, networks are defined as *a relatively loosely tied group of organizations that may comprise of members from government, university and industry who continuously collaborate to achieve innovation and TT.*

3.4. Success factors in NM

As the scope of this study focuses on the under-explored network level as discussed in Section 3.2.3, this study draws upon factors identified on a preliminary basis in the wider network literature for their impact upon the performance of whole networks rather than merely on the organization. As summarized in Table 4, these include structural, relational and cognitive factors (Inkpen and Tsang, 2005). The literature has discussed the impact of structural factors of centrality and density on network efficiencies such as communication efficiency (Oliver, 1991; Rowley, 1997). Similarly, the importance of relational factors, such as trust, on network efficiencies has also been identified in several studies (Powell, 1990; Rowley et al., 2000). Cognitive factors result in shared understanding, for example coordination, and have also been recognized for their influence on network outcomes (Denize et al., 2005; Inkpen and Tsang, 2005; Moenaert et al., 2000).

Table 4. Network success factors

Categories	Network Success Factors	Authors	Methodology Analysis	Future work
Network Efficiency	Communication Efficiency	(Ford and Johnsen, 2000; Ford and Johnsen, 2001; Huhtinen and Virolainen, 2002; Jung, 1980; Moenaert et al., 2000)	Studies have been descriptive and conceptual providing little empirical evidence.	Empirical testing is required.
Cognitive Factors	Coordination	(Achrol and Kotler, 1999; Guiltinan et al., 1980; McCosh et al., 1998; Medlin, 2006; Mohr et al., 1996; Moller et al., 2002; Ojasalo, 2004; Ruekert and Walker, 1987; Van de Ven and Walker, 1984; Van de Ven, 1976)	Limited empirical evidence has been provided based on qualitative case study. These studies adopt the view of a focal company.	An overall network view should be provided.
	Harmony	(Ford and Johnsen, 2000; Freeman, 2001; Gupta et al., 1986; Laine, 2002; Song and Thieme, 2006; Vaaland, 2001; Welch and Wilkinson, 2005)	Some studies have been descriptive and conceptual. Others have provided empirical evidence based on qualitative, case studies. However, those have adopted a relationship or focal company perspective or inter-functional perspective within companies.	An overall network view should be provided.
	Role Expectations	(Anderson et al., 1998; Biddle and Thomas, 1966; Chonko et al., 1986; Ford et al., 1975; Havila, 1992; Heikkinen et al., 2006; Knight and Harland, 2005; Järvelin and Mittilä, 2001; Minzberg, 1980; Mittilä, 2002; Moller et al., 2002; Montgomery, 1998; Netemeyer et al., 1996; Rizzo et al., 1970; Singh and Rhoads, 1991; Zurcher, 1983)	Some studies have identified specific roles that are important in networks. However, these are incomprehensive and context specific and thus not generalizable. No causation to overall network outcomes has been established.	Empirical evidence on the link between role expectations and network outcomes is required.
Structural Factor	Power Distribution	(Dahl, 1957; Dwyer, 1980; Frazier and Rody, 1991; Gaski, 1984; Hakansson and Vaaland, 2000; Lusch, 1976; Lusch and Brown, 1982; Medlin and Tornroos, 2006a; Sutton-Brady, 2000; Welch and Wilkinson, 2005; Wilkinson et al., 2004; Zolkiewski, 2001)	Most studies have been descriptive and conceptual. Few have provided empirical evidence based on qualitative, case studies. However, those have adopted relationship or focal company perspectives and have not established causality to network outcomes.	Analysis on the network level should be taken which establishes causation between power distribution and network outcomes.
Relational Factors	Trust	(Aulakh et al., 1996; Coote et al., 2003; Doney and Cannon, 1997; Ganesan, 1994; Morgan and Hunt, 1994Nooteboom, 1997 #450; Norman, 2002; Plewa, 2005)	Most studies have adopted the organizational or individual perspectives or at best the dyadic perspective.	Analysis at the network level should be undertaken.

33

3.4.1. Key factors for NM: Coordination, Harmony and Communication Efficiency

Coordination

Coordination has an established history in management research. Fayol's (1949) well-known definition of management also identified coordination as one of its five key elements. In traditional management with an intra-organizational focus, coordination is associated with organizing, planning and control and results in increased structure, hierarchization and organizational growth (Axelsson and Easton, 1992).

In addition to the traditional management research, coordination has been explored in an inter-organizational context in supply chain management (SCM) research. Coordination in SCM research is usually analyzed on the relationship level, such as coordination between retailers and their manufacturers as indicated in Mohr et al. (1996). In the SCM context, coordination refers to the organizing of network activities and relationships to improve activity cycle effectiveness (Axelsson and Easton, 1992). However, this view of coordination is more applicable to relationships within distribution networks rather than a holistic network perspective and in particular one that is pertinent to a TT or innovation context.

From a network perspective, the impact of coordination on network effectiveness is debatable. On one hand, in keeping with Ford et al.'s (2002) view, a network has no hub and cannot be controlled. The purpose of networks includes reducing hierarchies (Achrol and Kotler, 1999). On the other hand, coordination is necessary to ensure that multiple actors can work cohesively (McCosh et al., 1998). Empirical evidence has also been contradictory. A study by Ojasalo (2004) revealed that although actors in a network do not like hierarchies, they would like an actor who has the highest authority and responsibility to ensure that outcomes are achieved. He argues that a coordinator may be necessary who adopts a different role to traditional management that is characterized by hierarchies, opportunism and bureaucracy. This actor is sometimes described as the 'network captain' (Campbell and Wilson, 1996). Networks require a form of hybrid coordination between market and hierarchy. They have less administrative controls compared to hierarchies and more incentives as all actors should benefit from partaking in the network (Powell, 1990; Williamson, 1991). Studies on coordination have been undertaken on the level of analysis of the relationship or focal organization, and thus, an overall network approach is required (Moller et al., 2002). Medlin (2006) has highlighted the need for future research in network coordination since this area lacks empirical evidence.

Harmony – Conflict/ Cooperation

In addition to coordination, the level of harmony in the network may also impact on its outcomes. The marketing literature contains numerous studies on conflict and more specifically, the IMP literature contains studies on conflict and cooperation. These studies have been mainly dyadic (Welch and Wilkinson, 2005). Harmony is a term used in the new product development (NPD) literature that can be applied to both the NM and TT literature. The term encompasses both the notions of conflict and cooperation that have emerged in the NM literature but retains a more positive connotation than conflict for management. The NPD literature on harmony predominantly adopts an intra-organizational focus (mainly on the relationship between internal R&D and marketing functions) rather than one on inter-organizational collaborations. Thus, exploring harmony in an inter-organizational network context should contribute to the definition of this construct. Harmony is reflected in whether actors are 'involved from the early phases of the innovation, if they attempt to understand each other's point of view, if conflicts between them are resolved at the lowest possible level ... and if they discuss issues rather than simply accept them' (Gupta et al., 1986, p 12).

Several authors have tapped into Confucian writings in their discussion of harmony. Xie et al. (1998) argue that contrary to popular belief, the Confucian notion of harmony includes disagreement, diversity of opinions and open debates:

> The gentleman agrees with others without being an echo. The small man echoes without being in agreement. (Confucius cited in Lau, 1983).

Inherent in this quote is that morally superior people are able to maintain harmonious relationships even though they may have differing views. Small-minded people, on the other hand, echo another's opinions while secretly disagreeing, which is not real harmony (Xie et al., 1998). Therefore, the harmony construct should reflect that concave relationship (Xie et al., 1998) whereby measures for a moderate level of harmony should be described.

The term harmony is appropriate in capturing the synonymous, varying aspects of conflict and cooperation that have emerged in the NM literature. A degree of conflict may be required for innovation while at the same time cooperation may be needed for efficiency (Vaaland, 2001). Therefore, both collaboration and conflict may be necessary for innovation networks (Laine, 2002). However, the IMP studies on conflict and cooperation have been generally descriptive and conceptual and few have provided empirical evidence on the impact of conflict and cooperation on network outcomes. These studies have adopted a relationship or focal company perspective. A network approach is required to investigate the impact of both conflict and cooperation on network efficiency and effectiveness.

Communication Efficiency

Communication efficiency is an important success factor (Ford and Johnsen, 2000; Ford and Johnsen, 2001). Although the actor-resource-activity model alludes to the impact of network factors on communication efficiency (Gadde and Hakansson, 2001), no empirical evidence has been provided in support of this notion. Moenaert et al. (2000) argue that communication efficiency is a measure of communication effectiveness given its costs. They argue that for effectiveness to be achieved there must be motivation to share information. The transferor must be able and willing to transfer information (Jung, 1980) that could have an impact on the recipient. Effectiveness requirements include transparency of the communication network, knowledge codification and knowledge credibility. Efficiency requirements include cost of communication and secrecy (Moenaert et al., 2000).

3.4.2. The Interrelationships among Coordination, Harmony and Communication Efficiency

Coordination in the network may impact upon the level of harmony. Coordination is necessary to ensure that multiple actors could work cohesively (McCosh et al., 1998). Coordination may involve a level of formalization, clear definition of deliverables and a single authority who serves as a network manager. These factors may reduce the likelihood of escalation of conflict to unmanageable levels. Thus, harmony may be maintained.

In turn, the level of harmony may impact on communication efficiency. As explained in Section 3.4.1, harmony involves give-and-take in the relationships with both parties trying to understand the others' view points, incorporating them in early stages when setting the research agenda. Therefore, it is likely that these measures may increase communication efficiency in the network. Song et al. (2006) establish a link between harmony and the information gap as the latter can be a symptom of a lack of communication efficiency. The information gap is the difference between ideal and achieved levels of information sharing among participants (Song and Thieme, 2006, p 314). Information exchange is an aspect of communication efficiency (Denize et al., 2005; Moenaert et al., 2000).

Similarly, coordination may impact on communication efficiency in the network. Coordination may impact on the level of transparency, credibility and shared understanding in the network, which are key dimensions of communication efficiency as indicated by Moenaert et al. (2000).

3.4.3. Antecedents to Coordination, Harmony and Communication Efficiency

Power Distribution

Power distribution may influence coordination, harmony and communication efficiency in the network. The network literature indicates that power distribution of a network may be important for its effective management and in particular its coordination (Hakansson and Johanson, 1992 cited in Zolkiewski, 2001). Power in the network may also affect the level of conflict (or harmony) (Gaski, 1984; Lusch, 1976; Vaaland, 2001) and communication efficiency (Rowley et al., 2000; Rowley, 1997).

Although the study of power in marketing channels has had a long history, going back to the 1960s, the corresponding theoretical and empirical research has been predominantly dyadic rather than network-based (Wilkinson et al., 2004). Power is usually defined as the ability of one actor to control another and it stems from dependence in the dyad (Gaski, 1984; Hunt and John, 1974; Frazier and Rody, 1991; Lusch, 1976; Lusch and Brown, 1982; Welch and Wilkinson, 2005). Emerson's perspective on power can be found in many studies on marketing channels – 'channel member A's power over member B is directly related to B's dependence on A for scarce resources' (Dwyer, 1980, p 46). 'A has power over B to the extent that A can get B to do something that B would not otherwise do' (Dahl, 1957).

More recently, the need to explore the impact of power at the network level rather than at the dyadic level has been recognized (Welch and Wilkinson, 2005). The study done by Hadjikhani and Hakansson (1996) was significant in recognizing the impact of influences outside of a dyad on the behaviour of actors. Welch et al. (2005) also argue that network theories offer different perspectives on inter-firm power and advocate the need to further explore the power structure of the network in which a firm is embedded. While power can be gained through other avenues, such as resource attributes of individual actors, network theorists provide a complementary analysis of the structural sources of power (Brass, 1984).

Centrality and density are two such structural factors that have been recognized as important in the exchange of knowledge in the networks (Rowley et al., 2000; Rowley, 1997). Centrality is a measure of an actor's power derived from its network position (Brass and Burkhardt, 1993). Central actors can control and manipulate information exchanges between actors and impact on the communication efficiency in the network. One or two very powerful players may even drive some networks (Charles and Howells, 1992), whereas power distribution in others may be relatively balanced.

'Density is a characteristic of the whole network; it measures the relative number of ties in the network that link actors together and is calculated as a ratio of the number of relationships that exist in the network, compared with the total number of possible ties if each network member were tied to every other member' (Rowley, 1997, p 896). Density affects the level of power that any particular actor may exercise as highly interconnected networks with increased levels of information exchange may shape the power distribution. It also influences the level of coordination and communication efficiency (Achrol and Kotler, 1999; Oliver, 1991).

Many network studies on power have been descriptive and conceptual. Others have provided empirical evidence based on qualitative case studies (Hadjikhani and Hakansson, 1996). However, they have adopted relationship (Sutton-Brady, 2000) or focal company perspectives and have not established causality in relation to network outcomes. Therefore, further research is required to provide an analysis of the network level in order to establish causality between power distribution and network outcomes. Medlin (2006) also reiterates this call for future research about power in industrial networks.

Role Expectations

Role expectations in the network may also impact on the level of coordination required (Moller and Halinen, 1999; Heikkinen et al., 2007). Bengtsson et al. (2003) propose that 'informal agreements concerning an activity give rise to unclear roles and the conflict between different and unclear roles are more difficult to research than conflicts between clear roles' (Bengtsson et al., 2003, p 8). They also argue that the expectations of network actors may be different and contradictory. Role expectations generally involve the beliefs and attitudes about contributions and performance that actors hold of each other in a network (Heikkinen et al., 2006).

Despite the prominence of the role concept, its exploration at the network level of analysis remains limited. The concept of role has long been prevalent in the social sciences (Biddle and Thomas, 1966; Zurcher, 1983). It has also been applied in management research where various roles have been identified (Minzberg, 1980; Shenkar et al., 2004; Vilkinas and Cartan, 2001). Mitzberg (1980) describes various interpersonal (figurehead, leader and liaison), informational (monitor, disseminator

38

and spokesman) and decisional roles (entrepreneur, disturbance handler, resource allocator and negotiator). However, Heikkennen at al. (2007) argue that although these studies notably have some roles acting in an inter-organizational context e.g. Mitzberg's (1980) liaison, monitor and spokesperson, they generally focus on managing an organization. Similarly, marketing studies have applied concepts of role ambiguity, clarity and conflict particularly in researching salespeople but have focused on the organization or individual rather than the network context (Chonko et al., 1986; Ford et al., 1975; Netemeyer et al., 1996; Rizzo et al., 1970; Singh and Rhoads, 1991).

Although more recent studies have applied the concept of role in a network context, they have not attempted to link roles with outcomes of the network (Anderson et al., 1998; Havila, 1992; Heikkinen et al., 2006; Knight and Harland, 2005; Montgomery, 1998). Snow et al. (1992) propose three roles specific to the network context, namely, the architect, lead operator and caretaker, while Knight and Harland (2005) recommend six roles of advisor, information broker, network structuring agent, innovation facilitator, coordinator and supply policy maker/implementer. Additionally, Heikkenen et al. (2007) suggest roles appropriate for specifically managing in nets, including, webber, producer, facilitator, gatekeeper, entrant, aspirant, instigator, planner, compromiser, advocate, auxiliary and accessory provider. However, they acknowledge their limitation in not linking these roles to network outcomes to determine which roles are most beneficial to management and for theory development. Additionally, roles are context-specific and generalizability to other networks may be difficult (Knight and Harland, 2005). Nevertheless, attempts to measure and compare the importance of these roles for achieving network outcomes may be useful in NM.

Trust

Although studies on inter-organizational relationships have consistently established the importance of trust for relationship performance since the early 1990s (Seppanen et al., 2007), analysis at the network level remains limited. Influenced by prior research on trust from predominantly the socio-psychology literature and to a lesser degree the transaction cost economics literature (Sako and Helper, 1998; Young-Ybarra and Wiersema, 1999), the study of trust has generally moved from a level of analysis of individuals to organizations (Medlin and Quester, 2002; Seppanen et al., 2007).

In business studies and particularly in the field of marketing, trust is featured in numerous studies on business-to-business/ relationship marketing, sales management and channel management (Doney and Cannon, 1997; Ganesan, 1994; Morgan and Hunt, 1994). However, the network level of analysis remains under-explored empirically with extant studies focusing predominantly on organizational and even individual levels of analysis with one type of informant such as CEOs (Aulakh et al., 1996; Coote et al., 2003; Norman, 2002), salespersons (Ganesan, 1994; Nooteboom et al., 1997; Smith and Barclay, 1997), buyers (Mollering, 2002; Plank et al., 1999) and purchasers (Chow and Holden, 1997; Doney and Cannon, 1997; Zaheer et al., 1998) or at best the dyad e.g. universities and businesses (Plewa, 2005).

Despite the limited empirical studies of trust at the network level of analysis, network theorists have emphasized the importance of trust to network success, though conjecturally (Cravens et al., 1994). Several authors argue that trust influences network coordination as it is seen as network governance mechanism where networks with higher trust levels require less coordination and involve reduced governance costs (Powell, 1990; Rowley et al., 2000; Bidault and Jarillo, 1997; Seppanen et al., 2007). Others suggest that trust impacts on harmony as it facilitates conflict management as trusting network actors may forego short-sighted goals, voice their views openly and focus on developing shared initiatives (Achrol and Kotler, 1999; Powell, 1990; Rowley et al., 2000; Seppanen et al., 2007; Uzzi, 1996). Therefore, further research is necessary to provide empirical evidence of the impacts of trust on coordination and harmony at a network level of analysis.

3.5. Network Outcomes

In addition to utilizing measures of network efficiency, it is also important to evaluate their effectiveness. Ignoring overall effectiveness in achieving the objective of an initiative and simply focusing on efficiency is not appropriate in the innovation and TT process. Efficiency is seen as a necessary condition or hurdle, and effectiveness as the company's ability to generate a sustainable growth in its surrounding business network (Borgström, 2005). Therefore, measuring overall network effectiveness is important in determining network success.

While extant literature contains conceptual analyses of network effectiveness from multidimensional levels, measures of network level effectiveness remain under-developed (Sydow and Windeler, 1998). IMP authors have attempted to analyze network outcomes from the point of view of the organization, relationship and network as in Ford et al. (2001). However, clearly defined constructs and measures for network effectiveness have not yet been developed. Similarly, the management literature on inter-organizational networks contains a multi-layered perspective that is predominantly descriptive (Provan and Milward, 2001).

Network effectiveness

Most studies focus on analyzing organizational outcomes rather than network outcomes. Even the general network literature focuses on outcomes to the organization through network involvement, and mostly ignores issues of network-level effectiveness (Aldrich and Whetten, 1981; Knoke and Kuklinski, 1982; Marsden, 1990a; Provan and Milward, 1995). There is a need for developing a measure for network effectiveness. While an organization perspective may be useful for activities pertaining solely to one organization, the network perspective is fundamental for some scenarios (Provan and Milward, 1995). This is particularly true in cases of collaborative innovation where the successes of the initiatives are determined by the contribution of several players.

3.6. Summary

In order to advance understanding of the network perspective in TT, the NM literature was reviewed. Although no comprehensive NM model currently exist, several factors have been recognized as important in achieving network efficiencies and effectiveness as indicated in Figure 3. These include relational (trust), structural (power distribution) and cognitive factors (harmony, coordination and role expectation).

Figure 3. Conceptual Framework of NM

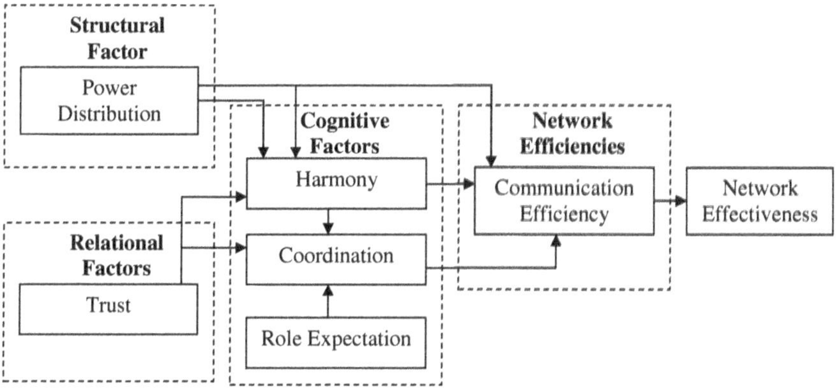

Following a review of the TT and NM literatures in chapters 2 and 3 respectively, chapter 4 discusses the model and hypothesis development.

4.0. Chapter Four - Qualitative Research Step: Model and Hypothesis Development

4.1. Overview

Research on the management of innovation networks surrounding TT is novel. Despite the international trend towards the creation of innovation policies, including grant systems to encourage inter-sectoral research involving TT, academic research in this field remains elementary (Corley et al., 2006; Plewa, 2005; Provan and Milward, 2001). As discussed in Chapter 2, research has surfaced in recent years that recognizes the significance of networks in innovation and TT. In order to develop these concepts further, the NM literature was reviewed in Chapter 3. However, empirical evidence on the important factors of NM has not been found. Therefore, this study develops a conceptual model that merges both TT and NM and contributes to theory development in both fields.

This chapter first justifies the research design that includes exploratory, descriptive and causal research. This study involves a multi-method approach using both qualitative and quantitative methods. As it combines NM and TT in a novel manner, qualitative research is useful in the exploratory stage to develop a conceptual framework and a set of hypotheses, and to serve as a basis for the design of the subsequent quantitative research. The quantitative research is useful in providing explanatory or causal evidence and in advancing theory in both the fields of NM and TT.

After providing an overview of the research design, this chapter describes and justifies the qualitative research methodology and findings. The qualitative research methodology adopts a case study approach in selected industries and involves in-depth interviews. The content of these interviews are analyzed using QSR NUD*IST N6, a software to increase the effectiveness of storing and exploring qualitative data (Richards, 2002). These findings are discussed and used to refine the conceptual framework that emerged from the literature and to develop related hypotheses.

4.2. Research Design

The research design is the framework for data collection and analysis (Cooper and Schindler, 2006; Ghauri and Gronhaug, 2005). It is the 'logic that links the data to be collected to the initial research questions' (Yin, 2003, p 19). The research design adopted in this study incorporates exploratory, descriptive and causal research. As illustrated in Figure 4, this study is based on two main phases of qualitative research followed by quantitative research. The former consists of exploratory research and the latter incorporates both descriptive and causal research. The research design could be best described as being an embedded, multiple case design (Ghauri and Gronhaug, 2005). This is because it comprises sub-units of measurement which facilitate empirical research (Yin, 2003). In this study, the sub-units are key informants within organizations partaking in the network. Multiple case studies are utilized as research is carried out within selected industries. Case studies can be used with the main types of research identified in the literature namely descriptive, exploratory and explanatory (Yin, 2003) and can also provide both qualitative and quantitative empirical evidence (Eisenhardt, 1989). They can offer a complementary approach and are justified in this study as it seeks to build a novel theory with frame breaking insights (Eisenhardt, 1989).

Figure 4. Flowchart illustrating research design

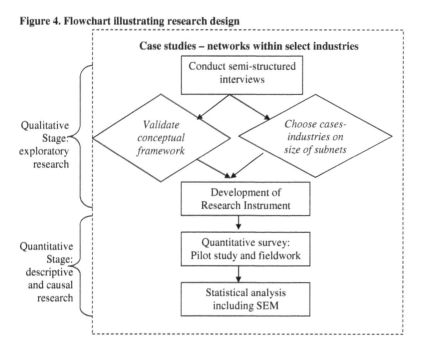

The exploratory stage used qualitative research methods and was required for a number of reasons. First, given the originality of the research, this stage was important in developing the conceptual framework. An exploration with key informants was useful in establishing important variables and relationships since the two fields of TT and NM are merged in a novel way in this study. Therefore, it was required for the development of hypotheses. It was also deemed essential in assessing the usefulness and relevance of the research to practitioners. Furthermore, it assists in determining the practicality of conducting formal research in identifying initial sub-networks within select industries as it involves carrying out interviews with key informants. In addition to contributing to the refinement of the conceptual framework and identifying the sampling units for formal research, it was also useful in developing the research instrument for further empirical research (Blaxter et al., 2001). Extant NM studies are based on little empirical research. Moreover, in the limited cases where empirical investigation has been used, the units of analyses remain the focal organization or, at best, the dyad (Provan and Milward, 1995). Similarly, existing operational definitions and constructs are biased towards the focal organization or dyad. Therefore, exploratory research is essential in developing the research instrument and the operational definitions and constructs which reflects the network level of analysis.

After the qualitative phase, quantitative research was conducted. Studies of networks in TT have either stressed the importance of combining both concepts without carrying out empirical analysis (Gibson et al., 1990) or in other cases have been limited to the perspective of the organization or dyad. Empirical research providing a network perspective of TT should advance theory development in both NM and TT and lead to greater generalizability of findings. Hence, both descriptive and explanatory quantitative research are important given the current lack of understanding on the subject and the acute need for NM theory development.

Descriptive research is also necessary given the lack of formalized studies on NM and networks in TT. Descriptive research leads to the determination of characteristics, proportions and associations between variables and involves calculations of means and variances (Cooper and Schindler, 2006). Therefore, it is necessary to provide structure and formality to the NM and TT network research streams conceptual findings currently predominates. Also, it serves as an intermediate foundation for further explanatory research.

Explanatory research is also important as it 'moves beyond description to seek to explain the patterns and trends observed' (Ticehurst and Veal, 2000, p 5). Establishing causality is important in theory development (Ghauri and Gronhaug, 2005). Therefore, explanatory or causal research is required when research aims to identify the key factors in managing innovation networks that lead to successful TT. Establishing causation increases the predictive ability of the theory and assists in providing useful management implications for innovation practitioners and policy makers. This study incorporates causal research by using Structural Equation Modelling (SEM) (Baumgartner and Homburg, 1996; Steenkamp and Baumgartner, 2000). It is based on a cross-sectional design which involves the measurement of cause and effect variables (Ghauri and Gronhaug, 2005).

4.3. Qualitative Research Methods

The qualitative research employed methods of case study analysis and in-depth interviews.

4.3.1. Case Studies

Qualitative research was used not only to explore and validate variables under investigation but also to select three industries and their networks based on network size and involvement in TT. Once these networks were identified, their participant organizations served as sampling units for the quantitative research to follow.

This study encompasses multi-method research within case studies. Yin (2003) argues that case studies can serve as a complementary and coordinated method that can be used simultaneously with qualitative and quantitative research. He goes on to justify the use of embedded case studies, one having sub-units of analysis, compared to holistic case studies where sampling sub-units are not identified. In this study, embedded case studies were used as the innovation network is the unit of analysis and its participating organizations are the sampling units. In this way, an overall network perspective was provided and is not limited to that of one particular organization while also focusing on NM processes of the wider inter-organizational network. This approach facilitates measurement of data and reduces the abstract nature of holistic cases (Yin, 2003). Finally, an overall case study approach is prevalent and justified in network research given the connected nature of networks (Iacobucci, 1996). Case studies are applicable as they are needed to explore the underlying NM processes under investigation.

45

Each case study was based on a different industry. The study investigated three industries which were chosen based on whether they contain networks of a sufficient size to subsequently support quantitative analysis. The pilot study was based on one industry. The full field work that follows limits its investigation to two different industries in order to provide an in-depth analysis of these industries. The two industries were deliberately selected to be different in nature to facilitate the identification of cross-sectoral patterns and trends.

4.3.2. Interviews

Semi-structured interviews were utilized within each case study. The topics, the general questions and their sequence were consistent throughout all interviews. However, probing into interviewees' thoughts was also used for discovery and exploration of new concepts (Cooper and Schindler, 2006). Each interview ranged from an hour to an hour and a half. Appendices B and C contain the information sheet and protocol used in the interviews. These interviews were necessary given the novelty of the area and the need to elicit rich information (Kumar, 1996). Their semi-structured nature was also justified over more structured interviews given the exploratory nature of the research and the need to refine variables identified in the literature and also to allow for the discovery of new variables. They were deemed preferable to unstructured interviews to facilitate analysis and to build on the conceptual framework that emerged from the literature in order to contribute to theory development.

Two phases of interviews were carried out. The first series of interviews aimed at choosing industries for further analysis and at developing a conceptual model. Interviewees were first asked to identify examples of innovation networks with which they were familiar. The key factors that the interviewees deemed important for successful NM and the manner in which they assessed the success of networks were then discussed. Towards the latter part of the interview, interviewees were shown the conceptual framework that emerged from the literature review. The relevance of its factors was discussed.

Dimensional quota sampling was used whereby key informants from each of four industries – biotechnology, ICT, automotive and wine were selected as well as TT/ commercialization specialists, as illustrated in Table 5. This type of sampling considers the significant dimensions of the population under investigation and chooses informants within each dimension, and thus, guarantees the inclusion of each dimension in the sample (Sarantakos, 1998). Representatives for each dimension were selected following the researcher's attendance at various commercialization, collaboration and TT events where industry specialists were present as well as a review of the websites, industry reports and annual reports of key government bodies and industry associations. Dimensional sampling was deemed necessary for interviewing key informants and TT specialists in each industry in order to gain an understanding of networks and to assess the practicality and relevance of the study.

Table 5. Dimensional quota sample

Industry or Category	Nature of Organization	Position	Title	Interviewe e #
Field Experts (Commercialization / TT / Networks)	Business	CEO	none	A1
	Quasi Government/ Private	Management	none	A2
	Government Agency	Management	none	A3
	University	Lecturer	Dr	A4
Biotechnology	Government Agency	Director	none	A5
	Research Centre	Deputy Director	Professor	A6
ICT/ Defence	Business	COO	none	A7
	Government Agency	Management	none	A8
Automotive	Research Centre	Management	Professor	A9
Wine	Research Centre	Technical Director	none	A10

The second series of interviews built on the findings from the first phase. At the end of the first series of interviews, a consensus was reached on the key success factors in NM and one industry was selected for a pilot study and two industries for full field work. The second series of interviews involved interviews with key informants from these three industries (see Table 6). These three industries consisted of networks of adequately large sizes to facilitate quantitative analysis. Given the novelty of NM research and the dearth of empirical work on networks, this second series of interviews was justified in order to refine the variables, further develop constructs and the research instrument for the proceeding quantitative research in a pilot study in one industry and full field work in two industries (Ticehurst and Veal, 2000). They were also essential in identifying the final sub-networks for quantitative analysis through snowballing within each industry by following up on referrals for the preliminary interview stage.

Table 6. Second Wave of Interviewees

Industry or Category	Nature of Organization	Position	Title	Interview #
Wine	University	Lecturer	Dr	B1
	Research Centre	Director	Prof	B2
	Business	Consultant	None	B3
Biotechnology	Research Centre	Director	None	B4
	Research Centre	Commercialization Manager	None	B5
	University	Lecturer	Prof	B6
Defence-related	Research Centre	Director	None	B7
ICT	Business	Director	Dr	B8

Interviews were recorded and subsequently transcribed with interviewees' consent. The recording of interviews produces a comprehensive set of verbatim comments (Ticehurst and Veal, 2000) that were useful in developing constructs, measures and definitions. Additionally, rather than being absorbed in excessive note taking, the interviewer could allocate time for building a rapport with the interviewee, establishing eye contact, assimilating the point of discussion and probing more effectively (Blaxter et al., 2001). Despite the shortcomings of inhibiting some respondents and unplanned technical failure of recording devices, the benefits of tape recording were deemed significant. To minimize the possible impact of this shortcoming, a few notes of the major points and contacts given were taken in the interview. This is seen as useful in having the crucial points recorded twice and also to demonstrate interest in the interview (Ghauri and Gronhaug, 2005).

The interviews were analyzed using QSR NUD*IST N6. Initially, nodes or key categories were developed based on the literature review but were altered during analysis to allow exploration of emerging topics. Findings from the analysis of the interviews were used to develop the conceptual framework and research instrument and to identify networks for quantitative analysis.

4.4. Conceptual Framework

This section discusses the preliminary results from the qualitative research, undertaken to refine the conceptual framework that emerged from the literature review into a causal model.

4.4.1. Coordination, Harmony, Communication and R&D efficiencies

Coordination

The qualitative research confirmed the contradictory findings in the literature that although too rigid formalization is a hindrance in network performance, some degree of it is required (Ojasalo, 2004). On the one hand, excessively rigid controls were seen as a barrier to creativity, as indicated by one of the interviewees:

> The results from scientists are based on experiments and serendipity so there can't be too many rigid controls. There must be a facility to allow for creativity. (Interviewee #A6)

Interviewee # B5 also felt that excessive reporting was a heavy burden.

> The centre is required to report to the government with high frequency and detail, so researchers have felt that they have spent a lot of time reporting on progress rather than conducting their experiments. There is a heavy burden of reporting. (Interviewee #B5)

On the other hand, a moderate level of formalization is also required. Some felt that specifications and contingencies should be clearly defined and that the collaboration should be explicitly verbalized, and discussed and written down in detail (Interviewees # A3, A8, A9 and B5).

> Contracting and formalizing the process makes the framework more durable. However, process related formalities such as standard operating procedures and teaming arrangements have not been widely adopted or effective. (Interviewee # A8)

The qualitative research also confirmed the need for a single coordinating authority to ensure the commitment and continuity of the collaboration. There should be an individual, group or organization either existing or new, taking responsibility for the collaboration and expected to take care of coordinating activities in the network and to exercise authority on behalf of the network if necessary. Some respondents felt that a new coordinating person or group might not be necessary for smaller or straightforward projects (Interviewees # A3 and A8). Nevertheless, in those cases, it was felt that an existing organization or group should be designated the responsibility to ensure that collaborators work in synchronization as shown in the following statement:

> Some collaborations fail because there is no particular actor committed to developing it. There is no active, single mind to develop it as a single initiative. Most parties are only focused on their own company's aims and objectives. Teaming is another model where there is one prime contractor and subcontractors bonded by teaming agreements. (Interviewee # A8)

49

Models for collaboration were found to vary. The coordinating group may not only comprise one authorized organization but an identifiable team of representatives from various organizations.

> It is standard to have a research management committee that has representatives of both organizations who reviews progress periodically. (Interviewee # B5)

Regardless of the type of coordinating mechanism, from a governing organization to a teaming arrangement, there should be a single coordinating authority to facilitate synchronization and professionalism in the network. 'It is essential for the successful coordination of the research network to identify specific contact persons / coordinators.... There is increasing recognition that the role of the overall or 'lead' coordinator within a network needs to be improved and made more professional' (Charles and Howells, 1992, p 168).

The qualitative evidence collected also confirmed the notion found in the literature that the role of the network manager is more than that of a coordinator compared to that of traditional management characterized by hierarchies, bureaucracy, centralism and opportunism. Some respondents felt that it was important for there to be someone who understands the capabilities, needs and expectations of parties and who could ensure synchronization. (Interviewees # A3, A5 and A9)

Harmony

Interviewees supported the notion that harmony is important in collaborative networks. In Section 3.4.1., harmony was described as reflecting whether actors are 'involved from the early phases of the innovation, if they attempt to understand each other's point of view, if conflicts between them are resolved at the lowest possible level ... and if they discuss issues rather than simply accept them' (Gupta et al., 1986, p 12). The interviews reiterated the need for both the research institution and commercialization partner to be involved in early phases when setting research agendas (Charles and Howells, 1992):

> A level of disharmony and compromise is involved between purely scientific and purely short-sited market driven agendas. For collaboration to be successful both the research institution and the commercialization partner should be involved early in the process and should have inputs into setting the research agenda. Both parties may be coming with different agendas and cultures but there should be open debate so that the agenda set will be workable and achieve common aims and be valuable for all parties. (Interviewee # A1)

Other interviewees (#B4 and B5) did not feel that conflict was a major issue. They felt that most of the conflicts were sorted out in the negotiation phase and that since the customer's needs were paramount, they should have the largest input in determining the direction of the project.

> The research agenda is negotiated before hand. The only tension is haggling about money. There may be commercial tension of agreement on how much a new invention is worth. Very typically academic institutions will think their technology is more valuable than industry will perceive it. Industry will say it's at a very early stage with high risk; we don't think it's worth 10 million dollars yet. Industry will think it will lead to new drugs and treatment. They expect to get a high return at a very early stage. Academics generally believe that technology is worth more that industry values it at. (Interviewee # B5)

The qualitative interviews also supported the need for addressing issues through discussion and establishing a common understanding rather than letting them escalate or simply accepting them.

The need for conflict resolution procedures was debatable. On the one hand, given the varying objectives of research institutions and industry partners, conflict resolution procedures are necessary (Interviewee # A3). On the other hand, it may not be necessary in certain subcontracting, highly formal cases as stated below.

> During the negotiation process, there is a need for open discussion and conflict management. However, once the contractual arrangements are made, conflict is not expected.
> (Interviewee # A9)

The interviewees who did not entirely support the need for conflict resolution mechanisms, did indicate that tensions and disagreements may be present even after the contractual arrangements and that there should be some system for addressing them. Therefore, the wording may have to be changed from conflict to tensions and disagreements as respondent may associate the word conflict with incidents and 'flare-ups' (Rosenberg and Stern, 1974). These latter two are a more advanced stage of the conflict process.

Communication Efficiency
The qualitative interviews confirmed the importance of communication efficiency. According to Moenaert (2000), such efficiency represents the effectiveness of communication, given its costs. He

argues that effectiveness requirements include transparency, codification and credibility. Efficiency requirements include cost and confidentiality. Interviewees stressed the importance of transparency:

> We made these projects public...in such a way that potential patents are not compromised and that it could inspire future collaborations ... We put it on a website and had an information day. (Interviewee # B2)

Interviewees also felt that knowledge codification was necessary for shared understanding.

> Persons from universities and industry talk in different languages. Translation must occur and both parties must be able to talk in a common language – rather than speaking French and English, they should begin speaking patois. (Interviewee # A3)

Additionally, the qualitative research also confirmed the need to address confidentiality issues via education of all partners:

> When an industry decides it wants to collaborate with a university group to fund research or as a partner on a linkage grant, there are commercial constraints that sometimes researchers are not used to - confidentiality, not being able to publish unless checking with the company, needing to make sure that intellectual property has been protected. So there is an education process. (Interviewee # B4)

Interviewees also discussed strategies for addressing issues of communication costs and confidentiality, such as housing collaborating organizations within the same building and applying intellectual property agreements to make the environment more 'membrane-like' rather than closed doors.

R&D Efficiency

The semi-structured interviews confirmed the importance and need to measure R&D efficiency given the diverse perspectives of actors operating in the network. Several interviewees expressed the challenge in finding a suitable measure that incorporates the varying views of collaborators. One interviewee alluded to the need for incorporating longer-term economic and social goals from a national perspective.

> Finding a suitable measure to assess TT success has been a challenge for government for a long time. We are currently in the process of developing a measure that also includes longer-term social and economic implications. If you can find a suitable inclusive measure, we will be

delighted because it is quite a task yet very important for future R&D investments. (Interviewee # A2)

Another interviewee reiterated the challenge in finding a holistic measure.

Measuring the success of TT has been challenging. We are now trying to adopt a measure that not only looks at patents and publications as has been traditionally done but one that incorporates other qualitative factors such as change in knowledge, attitudes, skills and aspirations of staff, the training of young researchers and other factors of the Bennett's hierarchy. (Interviewee # A10)

TT stakeholders have different perspectives of TT (McAdam et al., 2005). Consequently, the views of scientists in research organizations may vary from that of government agency sponsors providing infrastructure and funding or that of industry partners. The interviews reflected a diversity of objectives depending on the nature of the organization and interviewee's role. One interviewee from a quasi-government agency indicated that one form of evaluation of TT is the number of non-disclosure agreements signed, as his organization has a facilitating role in fostering relationships between research organizations and industry (Interviewee # A2). Additionally, this interviewee, and another from government, stressed the need to focus on other qualitative outcomes such as skill development, economic and social development. Yet another interviewee from a commercialization consultancy indicated that R&D expenditure is used as a measure of TT success as it indicates the organization's commitment to R&D that, in his opinion, would eventually lead to results. However, R&D expenditure does not imply that investments are sound nor that they would necessarily lead to results. One focus of R&D consultancy firms is to seek clients who spend on R&D, and thus, a measure that only focuses on one aspect of R&D inputs is inadequate given the broader objectives of TT actors.

Both of the scientists interviewed from research organizations stressed achievement of project milestones as the main measure of success. Therefore, the interviewees confirmed that there are a variety of measures used given different perspectives, and thus, a non-absolute, but a relative measure may be most appropriate in the network context.

A relative measure of R&D efficiency that gauges a perception of outputs given the inputs made by each actor is appropriate given the network context. The majority of the interviewees indicated that the compensation or outputs for each collaborator should be commensurate to their contributions made.

53

Therefore, the notion of R&D outputs in comparison to inputs reiterates the need to measure R&D efficiency.

> The compensation of each collaborator should be linked to each of their contributions made. In this way parties are fairly awarded for the value that they deliver. (Interviewee # A9)

4.4.2. Antecedents to Coordination and Harmony

Power Distribution

The preliminary interviews confirmed the importance of power in the collaborative network context to a certain degree. On the one hand, it came under question and a few interviewees sought clarification about this concept. One felt that power was not entirely applicable as collaborations may be comprised of both powerful and less powerful players who are all important once they contribute value in achieving the objectives of the collaboration (Interviewee # A6). Viewing networks as value nets was one suggestion by Moller (2001) which makes power less relevant.

On the other hand, other interviewees confirmed the impact of power which has been prominent in the marketing literature (Welch and Wilkinson, 2005; Zolkiewski, 2001). Interviewees # A3, A7, A8 and A9 indicated that when dealing with powerful players, formal subcontracting arrangements are established and there is less 'give-and-take' and input by subcontractors in setting the research agenda. Therefore, the centrality of a network may affect its dynamics, as suggested by Interviewee # A8.

> The level of give-and-take in network collaborations is determined by the power play. In cases where big players are involved, no questions are asked and work is simply executed by the others to the specifications set. (Interviewee # A8)

Similarly, another interviewee alluded to different types of power operating in the network, including economic power and expert power.

> Determining the research agenda does not cause problems: who pays the piper selects the tune. Companies' needs are paramount if they pay. Researchers get hearing because they are the experts and companies want to work with academic groups because they have expertise that the company does not have. (Interviewee # B5)

Although a unanimous acceptance of the importance of power per se was not established, interviewees did endorse its indirect influences on the relationship between trust and harmony.

Collaboration works where there is a shared vision and win-win approach, where one party wants to exploit the relationship, it doesn't work. Dispute resolution procedures, agreed exit strategies (harmony) can help, but, if one party wants to exploit the relationship (reduce trust), i.e. not mutual benefit, it won't work. Why would they want to exploit it? In rare cases it could be business strategy, in most cases, it's not. Arrogance and ego is usually the case there - suspect a conspiracy. It's more likely incompetence. (Interviewee # B8)

Trust

Although the focus of this study is not on the relational level but on the net level, a network is essentially a collection of relationships, and therefore, this study incorporates key relational dimensions as interviewees stressed that these contributed to the wider NM.

For collaboration to work it has to be treated like a partnership or relationship.
(Interviewee # B8)

Interviewees emphatically expressed that trust is fundamental in networks (Interviewees # A2, A3, A7, A10 and B7).

Building trust and building relationships is how we build networks (Interviewee # 7)

Additionally, the interviewees felt that trust influences the level of coordination required and the level of harmony in the network (Interviewees# A2 and A10).

Transfer Scope

Classification of the type of technological innovation to be transferred as either incremental or radical was problematic. Interviewees generally asked for clarification on this issue. Interviewee # A6 stated that there are no generalized categories for innovations based on the degree of novelty. Interviewee #B5 indicated that generally all research nowadays is incremental:

All research is incremental. Traditionally research was for the sake of new knowledge. More and more researchers have been asked to justify relevance of research and usefulness to society so there is a shift towards research more focused on outcomes that is useful to society. This is reflected in requirements for grants. There are few researchers that sit in ivory towers to do what they want to know rather than something that is useful. This had led to more researchers being involved in commercialisation as they can get more resources for their research in this way. (Interviewee # B5)

55

This resonates with recent findings from the literature which caution against revolutionary claims of the biotechnology industry, suggesting in retrospect that innovations have been incremental as they have built on previous models rather than being entirely new (Hopkins et al., 2007).

4.4.3. Outcomes

Network Effectiveness

The qualitative interviews confirmed the need to measure network effectiveness. Network effectiveness is particularly important for policy makers to ensure effective allocation of public funding (Provan and Milward, 2001) and investments made in ensuring that national priorities of improving innovative capacity and TT are met. It emerged from the interviews that little is known about the effectiveness of cooperative research centres (CRC). In Australia, CRCs comprise members of government, universities and industry. They are one of the major linkage mechanisms designed to improve Australia's innovative capacity. Therefore, research is required to link network characteristics and processes to network effectiveness in order to provide management implications. In this context, Interviewee # A9 states:

> To date, there is no study that I am aware of that assesses the effectiveness of the CRC model in Australia. Evaluating the effectiveness of these innovation networks is very important in ensuring that public funding is invested well. We are currently in the process of forming another CRC and implications from your research will be helpful in not only the present systems by also in developing policies that will guide the governing of future, similar collaborative arrangements even if by then they are referred to by a different name. (Interviewee # A9)

4.5. Conceptual Model

In order to develop a concise and applicable conceptual model, several refinements to the conceptual framework derived from the literature review were undertaken following the qualitative research.

First, the concept of role expectation was excluded from the model given its overlap with the measurable constructs of coordination and to a lesser degree harmony.

The qualitative research also highlighted the overlap between the constructs of coordination and role expectation (Interviewee # A3). Mohr et al. (1994) argue that coordination reflects the expectations of tasks that parties have of each other. In more recent research, Axelsson (2006) also combined role specification with coordination. The overlap with coordination is also obvious in some of the roles suggested in the literature such as webber (Heikkinen and Tahtinen, 2006; Heikkinen et al., 2007),

56

architect (Snow and Miles, 1992) and liaison (Minzberg, 1980). Similarly suggested roles in the extant literature also overlap with the construct of harmony, including the roles of compromiser and disturbance handler (Minzberg, 1980). Given that role theorists have not yet linked roles with network outcomes to establish the most important factors to NM theoretically and managerially (Heikkinen et al., 2007), attempting to establish the relationships between similar and yet measurable constructs of coordination and harmony may contribute towards theory advancement.

Second, transfer scope was eliminated from the model as it appeared problematic in a network context given the inability of interviewees to grasp the concept and to arrive at a consensus about it. Although various concepts for analyzing the scope of the technology are well established in the literature from an organizational level of analysis, application to a network level may not be straightforward (Cagliano et al., 2002).

Given the diversity of organizations involved in networks, their perceptions of the technology may vary. For example, two of the interviewees from research organizations (Interviewees # A6 and B5) viewed the somewhat 'radical innovation' as incremental as they have been working on related projects with this technology for several years. They argue that when applying for funding, they must show applicability and impact on business and market therefore, their research builds on previous research and is not radical research. However, government agencies and business partners may view the same technology as radical as these research projects are heavily funded given their novelty, risk and resulting anticipated market benefits and returns. Beckett (2005) argues that the perceived level of radicalness may be different based on the varying perspectives of the players involved. Understanding the perspectives and associated biases upon which these concepts were initially developed is important before applying them to a network context.

Even the use of other related innovation measures have come under criticism for being applied blindly without this consideration. Calaton et al. (2006) argue that it is important to distinguish the firm perspective from that of the customer when measuring disruptiveness. Conversely, Salomo et al. (2007) indiscriminately employ a multi-dimensional scale for innovativeness incorporating technology, market and internal and external factors to the firm by drawing on existing measures (Daneels and Kleinschmidt, 2001; Gatignon et al., 2002; Green et al., 1995). They conclude that product innovativeness does not influence the impact of management activities on innovation success. Hence, given the diversity of the perspectives involved in a network in terms of perceptions of the technology

57

with respect to the market, organizational goals and scientific field, reaching a consensus on transfer scope was not deemed relevant to a network context. As such the focus of this study was fine-tuned to addressing processes involved in NM, rather than incorporating the scope of the technology transferred.

Third, since there was very little consensus among interviewees on the impact of power distribution on harmony, this relationship was taken out of the model. While some interviewees did acknowledge that power distribution in a network affects its dynamics in terms of the level of negotiation and flexibility (harmony) (Kaltoft et al., 2005), others did feel that power was not a major factor. Irrespective of the size of the collaborator, they are valued based on their contribution to the initiative (Interviewee # A6). These networks become more equality-based rather than power-driven (Archrol, 1991). Cravens at al. (1994) argue that the role of power and dependence have diminished as these networks rely more on sharing and relational trust.

Nevertheless, the impact of power distribution on coordination will be retained. Power distribution via centrality may have a positive impact on coordination as highly central players can directly communicate with other players reducing the number of exchanges necessary through intermediary positions (Brass, 1984). However, it may also have a negative impact in reducing the negotiating power of weaker players (Bonacich, 1987). Similarly, Burt (1998) argues that density actually has a negative impact on performance as the ability to exploit unique information decreases with increased connectedness. Rowley at al. (2000) argue that in terms of governance, density is redundant in the presence of trust. Nevertheless, density may improve coordination which may both inherently improve with the number of ties (Oliver, 1991). Given the contradictions in the literature, the relationship between power distribution and coordination was retained for further investigation of its importance in the specific context of innovation networks.

Following the qualitative exploratory phase, the conceptual framework from the literature review was refined into a more concise and comprehensive model, as shown in Figure 5. It clearly identifies a number of hypotheses that represent current gaps in the literature, which this study now proposes to investigate empirically.

Figure 5. Revised Conceptual Model

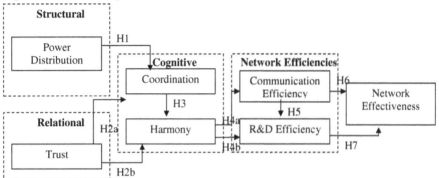

Hypotheses

The following hypotheses are directly derived from the conceptual model shown in Figure 5. They incorporate the relationships among structural, relational cognitive factors and their impacts on network efficiencies and overall effectiveness.

H1. Power distribution positively influences coordination.

H2a. Trust positively influences coordination.

H2b. Trust positively influences harmony.

H3. Coordination positively influences harmony.

H4a. Harmony positively influences communication efficiency.

H4b. Harmony positively influences R&D efficiency.

H5. Communication efficiency positively influences R&D efficiency.

H6. Communication efficiency positively influences network effectiveness.

H7. R&D efficiency positively influences network effectiveness.

4.6. Summary

This chapter first described the overall research design used for this study which consists of case studies and incorporates both qualitative and quantitative research. The chapter then justifies the qualitative methods used of case studies and semi-structured interviews. Case studies in three industries (wine, biotechnology and ICT) were chosen: The first for a pilot and the latter two to strengthen theory development by facilitating pattern matching. Two phases of interviews were presented with the former exploring and validating the variables and providing a preliminary identification of networks. The second stage was important in refining these networks and the research instrument to be used later in the quantitative stage. The findings from the qualitative research were subsequently discussed. The qualitative research also assisted in refining the conceptual model and developing the hypotheses. The methodology adopted to test the model will be discussed in chapter 5 and the results will be elaborated in chapter 6.

5.0. Chapter Five - Quantitative Research Step: Research Design

5.1. Overview

Given the value of multi-method research for developing and testing the conceptual model (Carson and Coviello, 1996), this study combines qualitative and quantitative research. Following the previous chapter which described the qualitative research methodology, this chapter discusses the quantitative research methodology. First, it describes the manner in which levels of theory, measurement and statistical analysis were dealt with and synchronized. This is important as the novelty of network level theory development and the abstract nature of networks warrant measurement at a subunit level when investigating network processes. The chapter then presents the selected industries and networks. The data collection method (survey via questionnaire) is discussed, including the strategies employed to ensure the respondents' common understanding of the network frame of reference. The pre-test used to fine-tune the wording and layout of the questionnaire is then detailed. Additionally, the operationalization of constructs is discussed which involves combining items from the existing literature with earlier findings from the qualitative research and then purifying and validating the developed scales via a pilot study.

5.2. The Levels of Theory, Measurement and Statistical Analysis

Given the novelty of NM empirical studies, it is critical to ensure the alignment between the level of theory at the network level and the levels of measurement and analysis. Extant theory has focused on the organization or relationship and the level of measurement has remained limited to either the focal organization or the dyad. Even in network studies, the network is generally the context rather than the target of the research and the corresponding theories that may result. Consequently, measures, constructs and operational definitions are tied to the organizational or relationship level rather than to the network level. In pioneering empirical research on NM and aiming towards network theory development, an articulation of these respective levels of theory, measurement and analysis is crucial. This is particularly important in improving the clarity, precision and rigour of the research, and subsequently, in reducing the likelihood of misinterpretation (Klein et al., 1994).

61

The level of theory in this study is the network level. 'The level of theory describes the target that a theorist or researcher aims to depict and explain. It is the level to which generalizations are made' (Rousseau 1985, p 4 cited in Klein et al., 1994, p 198). This study focuses on inter-organizational networks involved in innovation and TT and aims to explain success factors of managing such networks from a holistic perspective, not limited to the view of any one organization.

The level of measurement describes the actual source of the data – 'the unit to which data are directly attached' (Rousseau 1985, p 4 cited in Klein et al., 1994, p 198). When researching abstract phenomena lacking clear measures, it is sometimes useful to drill down to the subunit level of measurement while retaining focus at the higher level of analysis (Yin, 2003, p 45). The abstract nature of networks may account for the lack of empirical studies and theory development on NM and the bias towards organization specific or relational studies. In order to enable testability in this study, the key informants from network organizations were deemed the appropriate level of measurement. Multiple key informants within each organization were surveyed to improve the reliability of responses from each organization (Marsden, 1990b). These informants were focused on network rather than on intra-organizational issues. Retaining focus on the larger level of analysis while conducting measurement at the sub-unit level is an important device in focusing case study inquiry (Yin, 2003).

The unit of analysis of this study is the network. The level of statistical analysis describes the treatment of the data during statistical procedures.' (Klein et al., 1994, p 198). As multiple informants are surveyed, aggregation of these to the network level must be carried out to allow analysis at the network level.

In brief, the levels of theory and analysis are on the network while measurement is carried out by surveying multiple key informants within each organization while retaining focus on network issues. Given the abstract nature of networks, this is done to facilitate testability and tangibility. Though different in some respects, the levels of theory, measurement and analysis all return to the network level to enable conclusions to be drawn at that level.

5.3. Choice of Industries

Three industries were chosen: the wine industry was selected for the pilot study while the ICT and biotechnology/nanotechnology industries were employed for full fieldwork.

These industries were selected because of their paramount importance to Australia (ARC, 2008) as well as their international significance. The wine industry was chosen given Australia's increasing international prominence in this industry. The latter industries were selected as they are enabling technologies and have far-reaching implications in a range of related industries, thus increasing the potential impact and generalizability of this study.

The wine industry provides an exemplary industry in Australia. Interviewee # B6 explained that Australia's excellence in the wine industry is due to its roots in agriculture which has been a historical strength. This statement chimes with the statistics of Australia being the world's largest producer and forth largest exporter of wine (Austrade, 2007). Given its outstanding nature, key lessons could be learnt, and thus, it is an appropriate selection for the pilot. Additionally, the exploratory research revealed the presence of linkages among several organizations involved in collaboration and innovation, as illustrated in Figure 6.

Figure 6. Preliminary Sub-network in Wine Industry from First Stage of Interviews

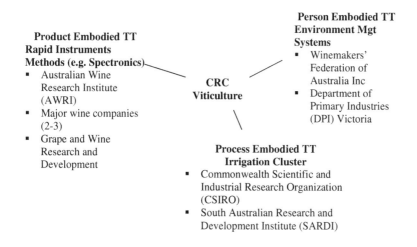

Two industries were selected for quantitative research. These were biotechnology/ nanotechnology and ICT (defence-related). The rationale for selecting these two contexts was not to facilitate direct comparison to reflect sharp variation but to identify trends that are relevant across contexts to improve explanatory value (Alvesson and Skoldberg, 2000; Swan et al., 2007). The degree to which emergent common trends could be generalized to other industries may be further magnified by the reach and impact of these enabling technologies on numerous industries (Charles and Howells, 1992). Several acronyms have been coined to reflect the anticipated impact of these transforming technologies which include NBIC (Nanotechnology, Biotechnology and Information and Communication Technology and Cognitive Science), and BANG (Bit, Atom, Neurone and Gene) (Bozeman et al., 2007, p 808). These industries were chosen due to their relatively large networks and their paramount importance both internationally and locally. ICT and biotechnology account for the majority of technology transfer from universities to industry and have experienced a shift from isolated pairs of collaborating organizations to dense, large networks comprising multitudes of organizations (Niosi, 2006; Roijakkers and Hagedoorn, 2006). Due to their pervasiveness, their networks are quite large and significant to many industries and they are, thus, deemed suitable for this study.

5.3.1. Biotechnology/Nanotechnology (B/N)

OECD defines biotechnology as 'the application of science and technology to living organisms, as well as parts, products and models thereof, to alter living or non-living materials for the production of knowledge, goods and services' (OECDb, 2006). It includes technologies dealing with DNA/RNA, proteins and other molecules, cell and tissue culture and engineering, process biotechnology techniques, gene and RNA vectors, bioinformatics and nano-biotechnology.

Biotechnology has transformed many industries. Powell (1996) argues that it has made vast impacts in the fields of pharmaceuticals, chemicals, agriculture, veterinary science, medicine and waste disposal. 'Biotechnology represents a competence-destroying innovation because it builds on a scientific basis (immunology and molecular biology) that differs significantly from the knowledge base (organic chemistry) of the more established pharmaceutical industry. Consequently, biotechnology generates enhanced research productivity with decreased risk, increased speed and potentially higher rewards' (Powell et al., 1996, p 117). Biotechnology is positioned to replace ICT as the leading strategic growth industry of the first half of the twenty first century (Ruttan, 2001). Figure 7 presents a sub-network in the biotechnology industry.

Figure 7. Preliminary Sub-network in Biotechnology Industry from First Stage of Interviews

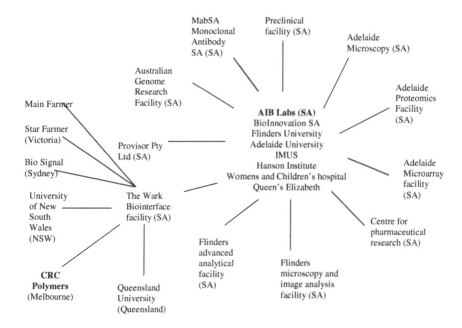

Networks comprise a fundamental part of the biotechnology industry. Over the last 20 years, the majority of TT, patents and spinoffs from academic research came from the biotechnology industry (Mowery et al., 2001). Network characteristics have also changed. Ruttan (2001) argues that prior to the mid-1970s, biotechnology research was encompassed predominantly within universities and research organizations with the focus on solving health problems. He states that in the late 1970s and early 1980s, the commercial potential became obvious and there was an intense formation of university-industry relationships and associated entrepreneurial activity. Compared to the 1980s, when small entrepreneurial biotechnology companies played a significant role in establishing relations between sub-networks, these networks are currently dominated by large multinationals (Mowery et al., 2001).

There is a degree of overlap between biotechnology and nanotechnology. Nanotechnology basically refers to the minute scale of components and the properties that they generate at that scale. Bozeman et al. (2007) argue that the US National Nanotechnology Initiative's definition as follows is fast becoming the standard definition:

> Nanotechnology is the understanding and control of matter at dimensions of roughly 1 to 100 nm, where unique phenomena enable novel application. The diameter of DNA, our genetic material, is in the 2.5 nm range, while red blood cells are approximately 2.5 µm. Encompassing nanoscale science, engineering and technology, nanotechnology involves imaging, measuring, modelling, and manipulating matter at this length scale. At the nanoscale, the physical, chemical, and biological properties of materials differ in fundamental and valuable ways from the properties of individual atoms and molecules or bulk matter. Nanotechnology R&D is directed towards understanding and creating improved materials, devices and systems that exploit these new properties. (Bozeman et al., 2007, p 808)

Indeed, the overlap between biotechnology and nanotechnology is apparent given the introduction in 2003 of a journal entirely devoted to nano-bioscience by the Institute of Electrical and Electronic Engineers (IEEE), an international, influential technology body, in addition to their separate journals on ICT, biotechnology and nanotechnology.

The junction between biotechnology and nanotechnology was chosen for further quantitative research rather than only the former for several reasons. Primarily, this cross-fertilized area is characterized by clear collaborations and networks given the necessity of research infrastructure.

In the initial wave of interviewees, Interviewee # A6 explained that revealing some of the partners involved in their strictly biotechnology collaborations was a breach of contractual obligations. This is because information on certain alliances between organizations is a source of competitive advantage as the products to be introduced become obvious given the competencies of players involved.

Other interviewees in the biotechnology industry were also apprehensive in identifying partners (Interviewees # B4 and B6). That said, Interviewee # A6 did inform that their organization was involved in biotechnology/nanotechnology research that involved research centres, facilities and networks where players were made public. This finding resonates with the literature that underlines the key role of research facilities as research at the nanoscale necessitates large investments from multiple organizations and therefore, requires the establishment of networks. These investments include large clean rooms, powerful microscopes and e-beam and nanoimprint lithography (Robinson et al., 2007).

Therefore, in contrast to the confidentiality concerns surfacing in the biotechnology industry, researching nanobiotechnology provides a viable platform for further quantitative research as collaborators could be revealed. The juncture between these two industries was also justified as advocates claim that their potential impact on related industries is greater given nanotechnology's relevance to both organic and non-organic materials compared to biotechnology's impact on predominantly organic materials (Rothaermel and Thursby, 2007). Therefore, eventual generalizability of findings may be increased by focusing on both.

5.3.2. Information and Communication Technologies (ICT)

Compared to still unrealized anticipated outcomes from later technological waves of biotechnology, and more recently nanotechnology (Bozeman et al., 2007; Ebers and Powell, 2007; Pisano, 2006), the impact of ICT has been felt more strongly in many related industries. These include defence, manufacturing, government, education, agriculture and health. OECD defines ICT as 'industries that support the electronic display, processing, storage and transmission of information' and included electronic equipment, telecommunications and computers (Jacques, 2002). The OECD definition includes:

> Manufacturing: 3000 – Office, accounting and computing machinery; 3130 – Insulated wire and cable; 3210 – Electronic valves and tubes and other electronic components; 3220 – Television and radio transmitters and apparatus for line telephony and line telegraphy; 3230 – Television and radio receiver, sound or video recording or reproducing apparatus and associated goods; 3312 – Instruments and appliances for measuring, checking, testing and navigating and other purposes, except industrial process equipment; 3313 – Industrial process equipment.

> Services: 5150 – Wholesaling of machinery, equipment and supplies (if possible only the wholesaling of ICT goods should be included); 7123 – Renting of office machinery and equipment (including computers); 6420 – Telecommunications; 72 – Computer related activities.
> (OECDa, 2006)

ICT was selected because of its significance to growth and innovation in many industries. Ninety percent of future innovation in the automotive industry will be driven by ICT and it represents 80% of costs of warships and submarines in the defence industry (EIA, 2007). Therefore, improving the management of innovation networks in ICT is important. The qualitative research revealed a number of institutions and linkages involving innovation and collaboration in the defence-related ICT industry in Australia as illustrated in Figure 8, and therefore, this industry was also selected for further quantitative analysis.

Figure 8. Preliminary Sub-network in ICT Industry from First Stage of Interviews

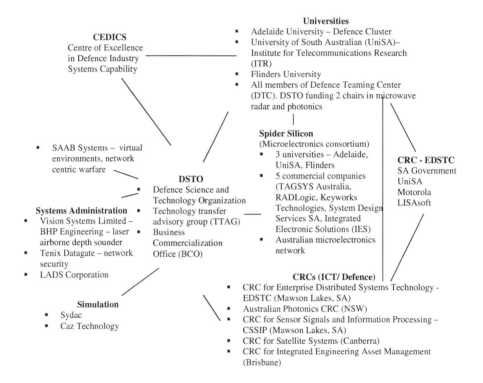

5.4. Data Collection Method

A questionnaire survey is an appropriate method to collect the data necessary to empirically test theoretical models. First, it facilitates the quantification of data (Ticehurst and Veal, 2000). Existing studies on NM are descriptive and conceptual in nature with little or no quantitative empirical evidence. Therefore, quantification would enhance theory development in this field. It also adds transparency and structure to the research in terms of data collection and analysis that may be replicated by other researchers (Ticehurst and Veal, 2000). Given the lack of constructs and measures at the network level, the constructs and measures proposed in this study should contribute usefully to the network stream. Furthermore, due to its ease of replication, the questionnaire survey can be applied to various industries, and thus, offer comparability in the methodology and an ability to demonstrate its reliability (Blaxter et al., 2001).

A hybrid survey combining a personally administered survey for initial respondents and then survey via mail to others within the network was deemed appropriate to reduce the disadvantages arising from survey research. The main disadvantages of survey research include respondents' potential misinterpretation of frame of reference, and low response and completion rates (Kinnear et al., 1996). Given the problematic nature of network boundaries and the intangible, abstract nature of networks, it was necessary to ensure an accurate identification of the network by a few key informants within the network. Therefore, for each sub-network, the survey was administered via personal interviews predominantly in order to ensure that the network was precisely identified until a consensus of core players was reached (Perry and Rao, 2007).

While interviewing using a questionnaire, Ticehurst and Veal (2000) argue that it is important that the precise wording of the questionnaire be used and in cases where the respondent may not understand the question, that it is simply repeated and then the next question asked if miscomprehension persists. Adhering to this protocol is crucial as further elaboration or explanation may lead to bias. Once a consensus on the network composition was reached, the remaining respondents within the sub-network can be surveyed via mail surveys, with a diagram of the agreed network composition included at the beginning of each questionnaire to clearly establish the relevant frame of reference (Kinnear et al., 1996). Surveying the remaining participants by mail was justified by the geographical dispersion of the respondents and costs and time constraints of the study (via post for the pilot study and via online survey for full fieldwork) (Cooper and Schindler, 2006). This hybrid method of recognition and recall improves reliability as it creates a common frame of reference while allowing flexibility in identifying new participating organizations with which there are common agreements (Cooper and Schindler, 2006; Marsden, 1990b; Wasserman and Faust, 1995). Previous network studies have identified problems when inaccurately determining the participants in a network, which can result in inconclusive results, and low and incomplete response rates. Therefore, this hybrid survey method aimed at reducing the likelihood of misrepresentation of the networks and improving response and completion rates.

5.5. Selection of Cases

Prior to the selection of final cases, attention has to be placed on identifying the networks by adopting various strategies of screening, focusing on 'issue-based nets' and snowballing.

Screening

Given the abstract nature of networks and the consequent misinterpretations which may result, screening is necessary to ensure that participants share a common agreement that they operate in a network. Screening increases the likelihood that cases are relevant to the phenomena under investigation, before proceeding into formal data collection (Yin, 2003).

Agreement on the network composition has been contentious in past network studies. As discussed in Section 5.4, this study adopts a hybrid administration of questionnaires whereby initial questionnaires for each sub-network were administered face-to-face in order to accurately identify the network and the later ones were undertaken by mail for the pilot study and online for the full fieldwork. During the face-to-face interviews, prior to network identification and questionnaire administration, screening was conducted. Furthermore, the network had to involve TT as some networks may be innovation related such as infrastructure development. Therefore, screening was also necessary to ensure the selection of networks where TT occurs.

Issue based nets

Given the boundary-less nature of networks, determining the absolute population or entire network at the industry level is challenging, if not impossible. This is the reason, according to Brito (1999), that empirical studies in the IMP literature are based on analysis at the dyadic level or from the focal organization perspective rather than the macro-network perspective. Other researchers also argue that empirical research and measurement of network processes and outcomes have been limited by the problematic nature of defining network boundaries (Mariko and Dodgson, 2003). However, this challenge can be overcome by investigating subsets of networks, issue-based 'nets', that can be identified based on their collaboration on specific issues. This was proposed by Brito (1999) as a solution as it provides an intermediary option between the extremes of boundary-less networks and focal organizations or dyads.

70

This study investigates sub-networks within larger industries that specifically involve collaborative innovation and TT. They are defined based on the perspectives of multiple key informants within those networks. The analysis of sub-networks is deemed justifiable as understanding those networks is useful from a management point of view compared to boundary-less networks that cannot be controlled.

Snowballing

Snowballing is used to identify the network through informants (Blaxter et al., 2001). Informants who are interviewed are asked to recommend others who meet the study criteria. They are interviewed and in turn are asked to recommend suitable informants (Sarantakos, 1998).

Connectedness is a central feature of networks and as such, methodologies based on random sampling and independence among units are not appropriate (Brito, 1999). Kenny and Judd (1986) rightly argue that 'the lack of independence is not simply a statistical nuisance that must be overcome, but rather, sometimes the interconnections are the very thing of interest in the research study (Iacobucci, 1996, p xiv). Therefore, snowballing is suitable given this connectedness.

5.6. Choice of Networks

This section describes the networks selected by applying social network analysis. It explains the choice of networks as being continuous by highlighting the duration of relationships and identifies the related industries of the networks under investigation as illustrated in Table 7.

Table 7. Characteristics of Final Respondents

Characteristics		Wine	ICT / Defence	Biotech / Nanotech
Number of Respondents (N)		51	124	95
Network size (number of organizations)		11	40	34
Density		0.3182	.0767	.0701
Centrality (network centralization)		41.48%	44.35%	41.94%
Duration of Relationships	0-2 years (N/%)	6 (8%)	11 (10%)	8 (9%)
	2-4 years (N/%)	12 (15%)	52 (46%)	40 (46%)
	4-6 years (N/%)	22 (28%)	30 (26%)	22 (26%)
	6+ years (N/%)	39 (49%)	21 (18%)	16 (19%)
Composition of Respondents	Business (N/%)	16 (28%)	15 (11%)	23 (22%)
	Government (N/%)	11(19%)	25 (18%)	16 (15%)
	University (N/%)	13 (23%)	70 (50%)	39 (38%)
	Research Organization (N/%)	17 (30%)	30 (21%)	26 (25%)
Related Industries (N)		Wine (50), Viticulture (2), Irrigation (1), Agriculture (1), Primary (1), Horticulture (1), Beverage (1), Education (1), Biotech (1)	ICT (13); Research (8); Chemical (8); Higher Ed (7); Polymer (6); Defence (5); Engineering (5); Nuclear Services (5); Manufacturing (3); Physics and R&D (3); Photonics (2); Agriculture (2); Interface Science (1); Electrical Cables (1); Health Science (1); Steel Manufacture (1); Solar (1); Water (1); Automotive (1); Hydrometallurgy (1); Neutron Scattering (1); Electron Microscopy (1)	Defence (22), Engineering (4), Radar(1), Forensic (1), Photonics (1), Robotics (1)

Network Description with Social Network Analysis

Social network analysis (SNA) was used to describe the size, centrality and density of the networks using the software Ucinet 8.0 (Borgatti et al., 1999). The size of the network used for the pilot study in the wine industry consists of 11 organizations compared to 40 for the ICT/ defence network and 34 for the biotechnology/ nanotechnology network.

Figures 9-11 contain diagrams of networks used for data collection.

Figure 9. Wine Network used in the Pilot Study

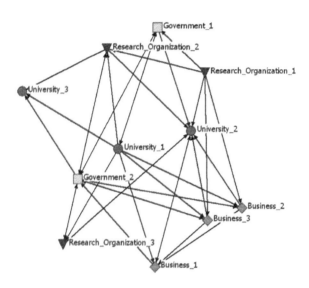

Key	#	Organization	#	Organization
University (U)	RO_1	Australian Wine Research Institute	G_1	NWA
Government (G)		Commonwealth Scientific and Industrial Research		Department of
Business (B)	RO_2	Organization	G_2	Primary Industries
Research		South Australian Research and		
Organization (RO)	RO_3	Development Institute	B_1	Orlando Wines
	U_1	University South Australia	B_2	Fosters Group
	U_2	Adelaide University	B_3	Hardy Wines
	U_3	Charles Sturt University		

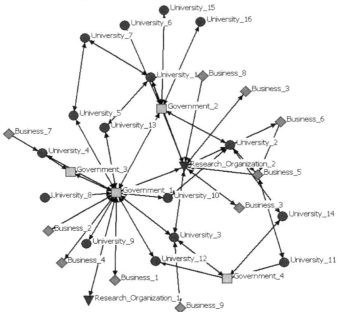

Key	#	Organization	#	Organization
	RO_1	Australian Stem Cell Centre	U_15	University of Western Australia
● University (U)	RO_2	CRC-Polymers	U_16	Queensland University of Tech.
□ Government (G)	U_1	University of New South Wales	B_1	Sola
◆ Business (B)	U_2	University of Sydney	B_2	Ceram Polymeric
	U_3	Monash University	B_3	Olex
▼ Research	U_4	Royal Melbourne Institute of Technology	B_3	Elisor
Organization (RO)	U_5	University Wollongong	B_4	Birchip Cropping Group
	U_6	Flinders University	B_5	Moldflow Pty. Ltd.
	U_7	University of South Australia	B_6	BlueScope Steel
	U_8	Curtin University	B_7	Lastek
	U_9	University of Queensland	B_8	Qenos
	U_10	Australian National University	B_9	Deakin
	U_11	Swinburne University	G_1	Commonwealth Scientific and Industrial Research Organization
	U_12	Griffith University	G_2	Australian Nuclear Science and Technology Organisation
	U_13	University of Technology Sydney	G_3	Nanotechnology Victoria Ltd
	U_14	Adelaide University	G_4	Defence Science and Technology Organisation

74

Figure 11. ICT Network used in Quantitative Fieldwork

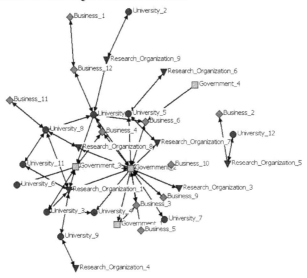

Key	#	Organization	#	Organization
● University (U)	RO_1	National ICT Australia	U_12	University of Western Australia
	RO_2	Kellar	U_13	Royal Melbourne Inst. of Tech.
▣ Government (G)	RO_3	CRC Sensor Signal and Info. Processing	B_1	AlphaJade
◆ Business (B)	RO_4	e-Health Research Centre	B_2	m.Net
▼ Research Organization (RO)	RO_5	Australian Nuclear Science and Technology Organisation	B_3	BAE
	RO_6	HxI Initiative	B_4	Tenix Systems
	RO_7	CRC Integrated Engineering Asset Management	B_5	Booz Allen Hamilton
	RO_8	Photonics CRC	B_6	Saab
	RO_9	CRC Smart Internet Technology	B_7	OSM Pty Ltd
	U_1	University of Adelaide	B_9	Thales Australia
	U_2	Swinburne University	B_10	SYDAC
	U_3	University of Melbourne	B_11	Moldflow Pty. Ltd.
	U_4	University of New South Wales at ADFA	B_12	MicroSoft
	U_5	University of South Australia	B_13	Galois Inc
	U_6	Queensland University of Technology	G_1	Rapid Prototyping,Development and Evaluation Program
	U_7	Institute for Telecommunications Research	G_2	Defence Science and Technology Organisation
	U_8	The University of Sydney	G_3	Commonwealth Scientific and Industrial Research Organization
	U_9	University of Queensland	G_5	Defence Material Organization
	U_11	Australian National University		

Continuous Networks

In keeping with the network definition provided in Section 3.3, empirical sampling of respondents was drawn from continuous networks. Ninety percent of the relationships in all three networks exceeded two years in duration. Additionally, fifty percent of relationships exceeded four years in the wine and biotechnology/nanotechnology networks while they accounted for over thirty percent in the ICT/ defence industry. Less that ten percent of all relationships were under two years, reflecting the continuous nature of the networks under investigation.

These findings are consistent with evidence from secondary sources such as annual reports, which identify the long standing sets of relationships in these networks. For example, the biotechnology/nanotechnology network included many organizations involved in the CRC-polymers, which is currently in its third funding cycle (CRC-Polymers, 2007). Each funding cycle is seven years, and therefore, many inter-organizational relationships exceed fourteen years (Monash, 2007). Similarly, relationships in the wine network are also continuous as some relationships span over five decades (AWRI, 2007; GWRDC, 2007). Likewise, collaborations between organizations in the ICT/ defence networks have also been longstanding and have been traced back thirty years among some participating organizations (DSTO, 2007).

Related Industries

Findings from analysis of these networks may be generalizable to other related industries given the impact of the selected networks on several other industries. As illustrated in Table 7, the ICT/ defence network overlaps with not only these two individual industries but also with radar, photonics and forensics. Similarly, the biotechnology/nanotechnology networks may have implications in related industries of not only biotechnology and nanotechnology but also materials, polymers, manufacturing, chemical, electrical cables, agriculture and nuclear services. According to Robinson et al. (2007), high outlays involved in acquiring nanotechnology infrastructure result in a number of related industries contributing towards the investment. These authors provide examples of two highly regarded areas for nanotechnology of Minatec in Grenoble, France, that focus on nanoelectonics (nanotechnology and electronics industries) whereas Twente in the Netherlands deal with materials and sensors. As such, the Australian network under investigation in Australia focus on a wide range of industries that also benefit from polymer and material research.

5.7. Questionnaire Design

5.7.1. Scales and Measurement

The majority of constructs in the questionnaire were operationalized using a multi-item, 7- point Likert scale. Different types of scales offer varied levels of measurement in terms of four characteristics – classification, order, distance and origin (Cooper and Schindler, 2006). The literature generally categorizes these scales into nominal, ordinal, interval and ratio scales. Nominal scales offer the lowest levels of measurement and assign numerical values to categories or codes (Ghauri and Gronhaug, 2005). They involve classification but do not reflect order, distance and natural origin (Cooper and Schindler, 2006). In this study, nominal scales are used for control variables such as the nature of institution and type, industry and location of office. In addition to classifying as nominal scales, ordinal scales rank categories using a qualitative scale. In comparison, interval scales adopt a quantitative scale with consistent distances between observations and an arbitrary origin. Unlike the latter, ratio scales have an absolute base point but also include a quantitative scale with consistent distances between points (Blaxter et al., 2001).

Likert scales, a type of ordinal scale (Kinnear et al., 1996), are mainly used in questionnaires. Using a multi-item, 7-point Likert scale for instance, users are asked to rate their views ranging from strongly agree to strongly disagree. An item refers to a measure that captures a specific attribute of a construct (Churchill, 1979). The use of Likert scales is justified as they are simple, easy to administer and rely on straightforward instructions (Kinnear et al., 1996). A 7- point scale is also justified as it increases the precision and sensitivity of the measures when compared with 5- or 3- point scales. Multi-item scales were also justified as they are often more reliable and less prone to measurement error, allow distinctions to be made amongst respondents and combine specific single measures, thus, reflecting more attributes of a construct (Churchill, 1979). Finally, the construct for network effectiveness was operationalized using a 10- point Juster scale, which is a type of interval scale. This is discussed further in Section 5.8 on construct operationalization.

5.7.2. Drafting of the Questionnaire and Pre-test

Questionnaire design addresses issues such as structure, wording and sequencing. In terms of structure, the questionnaire used in this study was divided into 4 main components. The first component was a request for co-operation via a cover letter. The second component consisted of instructions in the use and completion of the questionnaire and captured identification data regarding the respondents and classification data to indicate, for example, the industry. The third section sought to obtain data based on the network. It asked respondents to focus on network level issues such as power distribution, coordination and communication efficiency in the network as a whole. It contained a network diagram to increase clarification of the frame of reference (Kinnear et al., 1996). The frame of reference of innovation networks is important because in some cases respondents may associate these networks with a form of networking, i.e. socializing professionally to build business relationships. In other cases, they may associate it solely with informal arrangements which exclude more rigid forms such as contracting. Therefore, it was important to explain precisely that networks may comprise inter-organizational ties that vary in formality ranging from informal ones to contracting. Unlike the third section which focused on the network, the fourth section requested that respondents focus on a particular relationship and that they provide an assessment of trust. This was deemed necessary because trust in a network reflects the sum total of trust between any two of its actors. In this way, network level trust was derived based on dyadic relationships within the specific network.

In addition, further attention was given to using simple, clear language and to avoid jargon, ambiguity, double-barrelled questions and leading questions (Ticehurst and Veal, 2000). In terms of sequencing, questions were arranged in logical order with groupings between related topics.

The pre-test was an important stage in paving the way for the more formal quantitative assessments of validity (Cooper and Schindler, 2006; Kumar, 1996; Straub, 1989). Pre-testing refers to 'the initial testing of one or more aspects of the research design' (Kinnear et al., 1996, p 270). The use of the pre-test is important as it provides opportunities for: '(1) discovering ways to increase participant interest; (2) increasing the likelihood that participants will remain engaged to the completion of the survey, (3) discovering question content, wording, and sequencing problems, (4) discovering target question groups where researcher training may be needed, and (5) exploring ways to improve the overall quality of survey data' (Cooper and Schindler, 2006, p 385). Overall, pre-testing is useful in fine-tuning the structured research instrument (Kumar, 1996).

Using interviews, a pre-test was conducted within a selected network to increase the precision of findings and to purify the measures (Farrelly and Quester, 2003). Focusing on a specific network was important to ensure that a common frame of reference existed between respondents and researchers (Marsden, 1990). The selected issue-based network was focused on the issue of wine innovation. The composition of organizations in this network was determined by interviewing key informants from organizations within the network until a consensus of core players was reached (Perry and Rao, 2007).

78

These findings were also triangulated with information on collaborations from annual reports and public documents from various organizations in the network. To maintain the consistent frame of reference, a diagram of this network was included in the preliminary research instrument (Kinnear et al., 1996). Interviewees were also invited to comment on the items within the instrument to reduce ambiguities, and improve its design and flow.

Validity of the Pre-Tested Preliminary Research Instrument

Validity was adequately ensured in this preliminary research phase. Content validity is based on whether the domain of content is adequately sampled (Nunnally, 1970; Nunnally, 1978). Rather than testing this type of validity, Nunnally (1976) recommends several pro-active steps to ensure content validity through adequate selection of items from the domain and via appropriate procedures of construction. This study first ensured content validity by sampling items from the existing literature and incorporating them where relevant. It also adopted sensible procedures for construction: experts were included in the development of scales to determine whether they are representative and applicable (Garcia-Valderrama and Mulero-Mendigorri, 2005). Using experts in the selection of items also ensured face validity of measures. Construct validity was also addressed at this initial stage as the conceptualisation and operationalization of constructs were clarified and refined with experts (Straub, 1989). This is tested later through quantitative steps which are discussed in Chapter 6.

5.8. Operationalization of Constructs

This study adopts Churchill's (1979) procedure for developing measures. This procedure includes the following steps (1) specify the construct domain, (2) generate sample of items, (3) collect data, (4) purify measure, (5) collect data, (6) access reliability, (7) assess validity, and (8) develop norms. This chapter deals with the first four steps while the following chapter will address the remainder. The first two steps were carried out by tapping into the existing literature as well as incorporating the findings from the qualitative interviews. A list of measures for each construct was suggested by drawing on existing ones and providing justification where new ones were proposed.

Coordination

Coordination refers to the extent to which different parties in the relationship work well together in accomplishing a collective set of tasks (Mohr and Ravipreet, 1995; Van de Ven, 1976; Mohr et al.,

1996). No construct for network level coordination was found in the literature. Therefore, an attempt was made to develop this construct by extracting and altering relevant aspects from the coordination construct at the organizational and relationship levels and also by using the relevant findings from the exploratory interviews.

A construct for coordination at the organizational level developed by Ruekert (1987) adapted from Ven de Ven (1984), was not adequate for this research. The dimensions of this construct overlapped with other variables used in this research, such as communication, conflict and power. In that study, there were also overlaps in the dimensions of the 'conflict' and 'communication' with other variables used in the study. It was also placed in an intra-organizational rather than an inter-organizational, network context.

This existing coordination scale included a measure 'standard operating procedures (SOPs) have been established'. However, based on the findings from the exploratory interviews, it was found that process specifications such as teaming arrangements and SOPs are not common. However, the interviews highlighted the need for clearly defined deliverables. Therefore, there may be a need to distinguish between process and outcome formalization in the constructs and to include a measure relating to deliverables.

A study by Mohr et al. (1996) on supply chain management also included a construct for coordination (with manufacturer) that allows analysis at the relationship level. The scale items used were original to Guiltinan et al. (1980) and, when tested by Mohr at al. (1996), demonstrated acceptable convergent and discriminatory validity, based on confirmatory factor analysis (Bruner II et al., 2001). Therefore, these measures were incorporated and tailored to the network context.

In addition to altering measures used for coordination at the intra-organizational and relationship level, the qualitative research also alluded to other potential dimensions of coordination. For example, the need for a single authority and the coordinating role of the network manager were deemed important, and thus, relevant measures were included as a result. Table 8 details the items included in the scale for coordination.

Table 8. Coordination Scale

Deliverables were clearly defined.
The collaboration was written down in detail.
The collaboration was explicitly verbalized or discussed.
Our organization's programs were well-coordinated with the network's programs.
Our activities with this network were well coordinated.
We felt like we never knew what we are supposed to be doing for the collaboration.
We felt like we never knew when we were supposed to be contributing to the collaboration.
There was an individual, group or organization that took responsibility for the collaboration who was expected to take care of coordinating activities in the network and also exercising authority on behalf of the network if necessary.
A coordinating body was designated or identified that includes input from all collaborators.
A coordinating body ensured that all collaborators were working in synchronization.
The role of the 'network manager' was of more of a coordinator than of traditional management characterized by hierarchies, bureaucracy, centralization and opportunism.

Harmony

A construct for harmony can be found in the new product development (NPD) literature. A four-item scale is proposed by Song et al. (2006) adapted from Gupta (1986). Reliabilities for the harmony construct later used in the former study came under question and Song et al. (2006) called for a more robust measure of harmony. Xie (1998) argues that the operationalization of the harmony construct has to be expanded as extreme levels of harmony and disharmony may be counterproductive and a moderate, managed level of harmony may be suitable. Due to the focus on moderate or managed levels of conflict, dimensions reflecting conflict management mechanisms should be incorporated. Additionally, this construct is used to measure harmony between marketing and R&D functions internal to an organization, and thus, requires adaptation to the inter-organizational context. Furthermore, as indicated in the qualitative research reported in Chapter 4, the wording of 'conflict' was altered to 'tensions'. Additionally, as collaborations vary in formality, with parties interacting mostly during periods of formal or informal negotiation, the negotiation milestone was also incorporated. Table 9 details the items in the scale for harmony.

Table 9. Harmony Scale

During negotiation, meetings or discussions, there was give-and-take among participants. Each challenged the others and tried to understand the others points of view.
The research institution and the industry partner were involved in the early phases of discussion in setting the research agenda.
Conflicts between participants were resolved locally among the disagreeing participants rather than via escalation throughout the wider network.
There was compromise among participants in decision-making and each party obtained value from the network.
In the event of tensions or disagreement, an effective conflict resolution mechanism was in place.

Communication Efficiency

A measure for communication efficiency was not available in the literature. Nevertheless, a construct could be developed using Moenaert et al.'s (2000) definition of communication efficiency as a measure of communication effectiveness given its costs. According to these authors, (see Section 3.4.1), for effectiveness to be achieved there must be the motivation to share the information. The transferor must be able and willing to transfer the information that must have an impact on the recipient. Effectiveness requirements include transparency of the communication network, knowledge codification and knowledge credibility. Efficiency requirements include cost of communication and confidentiality. In addition to measures developed based on the aforementioned definition, the literature does offer measures of communication from which relevant components could be extracted as they related to communication difficulty that can be seen as a means of measuring efficiency (Ruekert 1987). Table 10 details the items in the scale for communication efficiency.

Table 10. Communication Efficiency Scale

The other participants were unable to transmit information that was required through the network.
The other participants were unwilling to transmit information that was expected through the network.
Information was not transmitted through the network because it was not valuable enough to be transmitted.
Information that we received via the collaboration lead to a change in knowledge.
Information that we received via the collaboration lead to a change in attitude.
Information that we received via the collaboration lead to a change in behaviour.
There were problems identifying the relevant persons to transfer information to or to obtain information from.
There was an understanding of the inputs made and progress of the collaboration.
Communication in the network was transparent.
Communication in the network was clear and accessible.
There were problems of knowledge credibility in the collaboration.
There were shared understanding of the meaning of information transferred among participants.

Information communicated by participants was not used.
Communication in the network was too costly.
There were no secrecy problems in the network.
There were no secrecy breaches in the network.
There were information leaks in the collaborative network.
When there was a need to communicate with other collaborators, there was difficulty in contacting them.
Difficulty was experienced in getting ideas clearly across to other collaborators when communication was made with them.

R&D Efficiency

R&D efficiency is a measure of R&D inputs relative to outputs. A measure for R&D efficiency can be found in the R&D econometrics literature. The Cobb-Douglas production function was used as a framework and R&D outputs are assessed in comparison to R&D inputs. The levels of analysis where this measure is used range from the firm's R&D department, to organizations, to regional innovation systems. R&D outputs are usually measured by identifying the number of patent applications filed in the last three years as in Fritsch (2000, 2001 and 2004). He measures R&D inputs as the levels of R&D expenditure made over the past three years. Additional specific variables are also used, such as ownership type and R&D infrastructure (Zhang et al., 2003), the Internet and knowledge spillovers (Kafouros, 2006). However, since the different actors in innovation networks contribute additional inputs other than those identified, such as skills and competencies, the measure used for R&D inputs needs to be more inclusive to cater for the network context. Similarly, the various actors have wider goals compared to number of patents arising, and therefore, the existing measure used for R&D outputs is inadequate for the purpose of this study.

As a result, a measure of R&D efficiency was developed by applying the fundamental concept and dimensions identified in the literature and expanded, based on the findings of the qualitative interviews. It is important that investments be made wisely regardless of their nature and that they provide fair returns to all players given their respective levels of inputs. Therefore, the general concept of R&D efficiency as a measure of R&D outputs to R&D inputs is relevant to the network context. Furthermore, a measure of R&D effectiveness that is relative and based on perception is useful in capturing the diversity of contributions and outcomes for all the actors involved. The literature revealed three main dimensions of R&D efficiency of cost, time and quality (Kafouros, 2006), and therefore, these would be incorporated into the development of the construct. The qualitative interviews also revealed another important dimension of equity in rewards whereby the value derived by collaborators is linked to their contributions. Table 11 details the items in the scale for R&D efficiency.

83

Table 11. R&D Efficiency Scale

The collaboration in the network was productive.
The collaboration generated sufficient outputs for the investments of resources.
The collaboration resulted in value for money.
The time spent in the collaboration was worthwhile.
We were always delighted with the performance coming out of this network.
The outcomes from the collaboration justified expenditures.
Compensation of participants was linked to their deliverables.
Participants in the collaboration obtained fair returns based on the value that they contributed.

Power Distribution

There is no consensus in the literature on the dimensions that may exist in measuring the phenomenon of power in networks and scales applicable in the network context have not been found, as existing measures are generally biased towards dyads. Zolkiewski (2001) argues that the lack of agreed dimensions is not surprising since many areas of management discuss power such as sociology, organizational change, economics and marketing. Similarly, the empirical assessment of power has been skewed. The generally accepted definition of power as the ability of one actor to influence another also reflects a dyadic perspective (Zolkiewski, 2001).

Some writers incorporate sources of power pertaining to reward, coerciveness, legitimacy, reference and expert power that stem from French and Ravern's (1959) study (Mohr and Nevin, 1990; Vaaland, 2001). Gaski (1984) provides a review of studies that analyzed power and found their validity to be questionable or their potential to establish causality to be minimal (Etgar, 1978; Hunt and John, 1974; Lusch, 1976; Lusch, 1977; Porter, 1974). These studies predominantly dealt with power in dyads rather than power distribution in a network: the latter being the focus of this study.

The concept of power distribution has been explored in a network context but this remains limited. Centrality is a structural measure of power distribution used in social network analysis. Cook (1983) argues that unlike power-dependence concepts that are dyadic, centrality takes the power structure of the entire network into consideration. However, some measures of centrality are egocentric as they examine the centrality of a particular focal organization within a network. These include degree, betweenness and closeness centrality (Freeman, 1979; Freeman et al., 1979). However, there are also network measures of centrality such as graph centrality, compactness or dominance in whole networks (Freeman, 1979; Snijders, 1981), which may be applicable for examination at the network level. These are described as macro-structural constructs, in contrast to their micro-structural counterparts (Kang, 2007).

Nevertheless, little is known about the reliability and validity of network measures (Marsden, 1990b). Furthermore, although mathematical and graphing techniques are used, they remain descriptive measures and have not been used in hypothesis testing to advance theory development (Cook et al., 1983). Therefore, further work is required to develop measures that are conducive to theory development.

In order to develop appropriate power distribution measures, the IMP work can be taken into consideration. Zolkiewski (2001) has identified dimensions of power analysis at the network level that may be useful in developing appropriate measures. These include the ability to influence the network and the ability to invoke political and media action. In the IMP literature, an actor's position in the network generally determines its level of power, as it influences its access to resources, control over key activities, relationships and ability to influence others. However, the IMP literature tends to be descriptive and conceptual and it does not provide appropriate network level constructs. Therefore, initial items used in scale development as detailed in Table 12, were drawn from concepts from the IMP literature including Zolkiewski's (2001) work, graph centrality literature (Freeman, 1979) and findings from the qualitative interviews.

Table 12. Power Distribution Scale

One or more large participants dominated the network.
The power distribution in the network was even.
My organization had the same amount of power as the other participants' organizations.
A particular participant had tremendous influence over the other players in the network.
A particular participant could have invoked political or media action.
A particular participant had more control than others over resources such as funding, equipment, skills and competencies.

Trust

Despite the lack of consensus on the operationalization of trust, measures of acceptable reliability and validity have been previously used in the TT context, and thus, will be used again in this study. Existing definitions and dimensions of trust vary. It may be defined as 'confidence in an exchange partner's reliability and integrity' (Morgan and Hunt, 1994, p 23). Others include dimensions of

85

benevolence and credibility in their definition (Doney and Cannon, 1997) while other definitions exclude them (Ganesan, 1994). Seppanen at al. (2007, p 255) argue that there is a lack of universal agreement on the number of dimensions which include 'credibility, benevolence, confidence, reliability, integrity, honesty, institutionalization, habitualization, ability, dependability, responsibility, likeability, judgement, goodwill trust, contractual trust, competence trust, fairness, reciprocity, togetherness, predictability, openness and frankness'. They also point out that there is disagreement on the semantic meaning of these words as capability, ability, competence and credibility are all used to refer to similar phenomenon while dependability, goodwill, benevolence, integrity and predictability all overlap.

Medlin and Quester (2002) argue that trust should be treated as a one-dimensional, global measure given the difficulties in conceptualization and operationalization associated with semantic ambiguity due to industry, national and cultural approaches (Seppanen et al., 2007). Plewa (2005) also treats it as a one-dimensional measure given the high correlation between dimensions. As the latter study achieves acceptable reliabilities and validities in a TT context, this study follows Plewa (2005) by applying measures from Doney and Canon (1997), Ganesan (1994) and Morgan and Hunt (1994) and slightly adjusts them for relevance in a network context as detailed in Table 13.

Table 13. Trust Scale

This partner kept promises it made to our organization.
This partner was not always honest with us.
We believed the information that this partner provided us.
The partner was genuinely concerned that our efforts succeeded.
When making important decisions, this partner considered our welfare as well as its own.
We trusted this partner to keep our best interests in mind.
This partner was trustworthy.
We found it necessary to be cautious with this partner.
This partner made sacrifices for us in the past.
We felt that this partner was on our side.
This partner was frank in dealing with us.
This partner could be counted on to do what is right.
In our relationship, this partner had high integrity.

Network Effectiveness

The difficulty in measuring network effectiveness is magnified not only by the diverse perspectives that may be present within any one organization but also by the multiplicity of perspectives of the groups of

86

organizations operating in a network. The literature on organizational effectiveness reinforces the view that assessment criteria should cater for the multiple perspectives existing in organizations (Cameron, 1986). Sydow (1998) argues that given the challenge in evaluating the effectiveness of organizations with clear boundaries, assessment of inter-organizational networks where boundaries may be unclear, is even more problematic.

To overcome this issue, a relationship marketing (RM) approach may be undertaken. There have been dyadic studies in the RM literature that incorporate outcome measures, given the seemingly varying views of two parties involved. Juster scales have been used to measure the outcome variable 'intention to renew' [the relationship] (Farrelly, 2002 cited in Plewa, 2005). As such, using a Juster scale ranking of 0% - 100%, allows respondent to rate their perceptions of network effectiveness. If the network is effective, it means it has worked positively for the organization within which a respondent operates. This was deemed appropriate, given the multiplicity of views in the network context which exceeds that of the relationship level of analysis. Furthermore, according to Sydow (1998) such a measure based on perception is appropriate as effectiveness remains subject to assessment by the other network firms. Table 14 illustrates the scale for network effectiveness.

Table 14. Network Effectiveness Scale

Using the following scale, in your view please indicate the level of effectiveness of this network. Please **circle one percentage** (%) figure. Zero percent (0%) indicates the lowest level of effectiveness and one hundred percent (100%) indicates the highest level of effectiveness.
0% 10% 20% 30% 40% 50% 60% 70% 80% 90% 100%

5.9. Statistical Analysis

Following the qualitative pre-test of the preliminary research instrument, a quantitative pilot study and a full field survey were conducted. The pilot study was undertaken in a network in the wine industry to purify and validate the developed measures, using exploratory factor analysis and internal reliability assessment (Ticehurst and Veal, 2000). Full field work was then carried out in ICT/defence and biotechnology/nanotechnology networks. At that stage, confirmatory factor analysis via structural equation modelling was used. Structural equation modelling is applicable in addressing the research

87

question of this study – 'what' are the key factors in managing collaborative innovative networks and 'how' do they lead to effective TT. These analyses will be useful in statistically highlighting the key factors and their impact on TT.

The results from the survey will not be pooled across industries as a case design requires that patterns be analyzed within each industry separately before they are compared (Yin, 2003). If trends emerge, it may be possible to generalize results beyond the two industries under analysis. Conversely, differences may offer industry-specific insights. In general, quantitative analysis is useful in providing evidence to advance theory development as both the TT and NM literatures lack quantitative empirical evidence from a network level of analysis. Finally, as the developed scales are a major contribution of this study, the scales will be tested for construct reliability, convergent and discriminant validity.

5.10. Pilot Study

Similar to other network studies, the pilot study was based on a small sample as all participants should be involved in the same network. This pilot survey was conducted from March to August 2007. It was based on 51 responses from a total of 108 persons surveyed representing a response rate of 47%. Respondents included 17 from research organizations, 11 from government, 16 from business and 13 from university. Given the small sample size, attempts were made to contact all non-respondents to determine reasons for non-participation. Four members of government and university cited possible confidentiality concerns and one director from a government agency surveyed cited conflict of interest and legal obligations binding directors. Additionally, three academics mentioned the increasingly demanding nature of their jobs and nine persons from wine companies were unreachable given recent mergers and subsequent reduction of staff. Five persons indicated that they were not suitable informants. Due to the importance of having suitable informants that collaborate together on a continuous basis within the same network, quantity was traded for appropriateness of respondents.

5.10.1. Scale Purification

Exploratory factor analysis and coefficient alpha were used to purify the measures (Churchill, 1979). Prior to this analysis, data preparation was carried out by recoding respective items and dealing with missing data using the method of estimation maximization, which has been identified as outperforming alternative methods, such as mean substitution or regression imputation which may understate variance

88

(Hair et al., 2006; Kline, 2005). Rotation was also carried out, using Varimax oblique rotation. This produces constructs which are more theoretically meaningful with improved distinctiveness of factors compared to orthogonal rotation or other forms of oblique rotation such as Quartimax, making it more appropriate given the scale development objective of this study (Hair et al., 2006).

Exploratory factor analysis using SPSS 13.0 was then undertaken to assess the structure of the variables and confirm dimensions discussed in the literature. Items which correlate highly, load well on factors, reflect broader dimensions (Hair et al., 2006). Loadings above 0.6 or even 0.5 are acceptable for early stages of exploratory research if other items pertaining to the same factor have higher loadings (Chin, 1998). Coefficient alpha, a widely used method for assessing reliability, was also calculated. Although there is little consensus on acceptable levels, Nunnally (1967) indicates that for early stages of exploratory research, reliabilities of 0.5 are acceptable. Following Churchill (1979), coefficient alpha was also calculated separately for each dimension and scale items were deleted based on whether they led to drastic reductions in coefficient alpha. This was determined by comparing their item-to-total correlations.

Table 15. Results from Scale Purification (wine industry network)

Constructs	Sub-dimensions	Coefficient Alpha	Original # of Items	# of Deleted Items	Final # of Items
Harmony		0.811	5	2	3
Coordination	Formalization synchronizing body	0.781 0.935	11	6	5
Power distribution		0.804	6	3	3
Communication efficiency	Transparency Codification Credibility Costs	0.861 0.855 0.810 0.568	19	7	12
R&D efficiency		0.912	8	2	6

Table 15 provides the results of scale purification. The pilot study confirmed the uni-dimensional nature of harmony. All items loaded well and yielded an acceptable coefficient alpha of 0.811 as illustrated in Table 15. The pilot study confirmed the two-dimensional nature of coordination discussed in the network literature and revealed during exploratory interviews. Although coordination should not be excessive, there should be adequate formalization and a coordinating body with a synchronizing role to ensure continuity (Ojasalo, 2004). Factor loadings for items within these two dimensions generally exceeded 0.7 and dimensional coefficient alphas exceeded 0.7.

The one-dimensional nature of structural power distribution was confirmed with all items loadings and coefficient alpha reaching acceptable levels. The factor analysis confirmed dimensions for communication efficiency of effectiveness - transparency, knowledge codification and credibility given costs and confidentiality issues (Moenaert et al., 2000). Dimensional coefficient alphas were above 0.5. The pilot study also confirmed the nature of R&D efficiency as a comparison of R&D outputs to inputs. All items loaded well and achieved a high coefficient alpha of 0.912.

5.10.2. Evaluation of Pilot Scales

Once the constructs were developed, they were evaluated using preliminary tests for reliability and validity. Table 16 illustrates the results. Reliability and validity are closely related through different conditions. Reliability addresses the precision or consistency of measurement while validity concerns with the accuracy and representation that a scale measures the intended variable (Nunnally, 1970). Therefore, while valid scales are reliable, the opposite is not necessarily true, and consequently, tests for both reliability and validity are required (Nunnally, 1967). Using AMOS 6.0, congeneric models for each separate construct were developed. These provided information on item loadings and error measurement that was essential in the calculation of reliabilities and validities (Byrne, 2001; Kline, 2005).

Table 16. Reliability and Variance Extracted for Constructs

	Construct Reliability	Variance Extracted
Harmony	.811	.590
Power Distribution	.814	.603
Coordination	.925	.750
Communication Efficiency	.937	.658
R&D Efficiency	.912	.635

5.10.3. Reliability of Pilot Scales

In addition to coefficient alpha, construct reliability was used to test reliability. Construct or composite reliability was calculated using information on standardised loadings of items and measurement error (ε_j) based on the following formula (Hair et al., 2006, p 642):

$$\text{Construct reliability} = \frac{(\sum \text{std. loading})^2}{(\sum \text{std. loading})^2 + \sum \varepsilon_j}$$

Convention suggests that construct reliability scores should exceed 0.5 (Hair et al., 2006) with those above 0.7 being desirable (Fornell and Larcker, 1981). As illustrated from the results shown in Table 16, all constructs achieved acceptable construct reliability ranging from 0.811 to 0.937.

5.10.4. Convergent Validity of Pilot Scales

Convergent validity assesses the extent to which measures of the same construct correlate (Churchill, 1979). It was calculated by examining variance extracted which is a measure of the amount of variance reflected in the construct relative to the variance lost to measurement error (Fornell and Larcker, 1981). The following formula is used to measure variance extracted (Hair et al., 2006, p 642):

$$\text{Variance extracted} = \frac{\sum \text{std. loading}^2}{\sum \text{std. loading}^2 + \sum \varepsilon_j}$$

It is recommended that variance extracted exceed 0.5 (Hair et al., 2006). If it is less than 0.5, this means that the construct captures less than 50% of the variance (Fornell and Larcker, 1981). As illustrated in Table 16, evidence of convergent validity was provided for all constructs as variance extracted was greater than the lower limit of 0.5.

5.10.5. Discriminant Validity of Pilot Scales

In contrast to convergent validity, discriminant validity measures the distinctness and novelty of each individual measure (Churchill, 1979). The main criterion to assess discriminant validity is for each item to load more highly on its specific factor than on any others (Moon and Kim, 2001; Yi et al., 2006). All items demonstrated acceptable discriminant validity as item loadings for their respective factors exceeded the required lower limit of 0.5 (Moon and Kim, 2001) as presented in Table 17.

Table 17. Item loadings for pilot scales

Factor	Loadings				
Power Distribution	.902 .906 .731				
Coordination		.746 .818 .881 .964 .942			
Harmony			.866 .846 .844		
Communication Efficiency				.741 .895 .704 .797 .779 .852 .695 .888 .872 .565 .545 .793	
R&D Efficiency					.830 .839 .818 .834 .842 .841

Therefore, the evaluation indicates that the developed measures demonstrate acceptable reliability and validity.

5.11. Post-pilot Changes to Questionnaire

The pilot study was useful in making certain changes to the questionnaire to be used in fieldwork. Changes were made to questionnaire length and the network diagram. However, all items were retained in the full survey as it was deemed premature for final deletion following analysis of the pilot data.

The first main change was a reduction in the length of the questionnaire. The pilot questionnaire included 9 pages as shown in Appendix C. It required respondents to comment on the network as well as four specific relationships. All respondents in the pilot study provided complete responses for the network and the first relationship. However, only six respondents out of a total of 52 commented on all four relationships. Therefore, the questionnaire was changed to capture data on one relationship. This was deemed sufficient since the study concerned mainly network level issues rather than dyads. In addition to the elimination of 3 pages which sourced data of relationships, the two introductory pages of the question were merged into one page which contributed to a further reduction in length. The reduction in length was useful in decreasing respondent fatigue and deterrence from completion due to length. The final questionnaire used in fieldwork is provided in Appendix D.

The second change to the questionnaire pertained to the network diagram. The diagram used in the pilot study provided the names of organizations in the network based on the findings from the qualitative research. However, it was found that this initial selection was limited and respondents suggested additional network participants. In order for the diagram to be more inclusive and ease visualization and understanding of the network concept as intended, a more inclusive diagram is required. Consequently, the questionnaire was changed after the pilot study to identify general categories rather than names of organizations. Categories was labelled was universities, businesses, government and research organizations. A couple names of organizations were only provided as examples to these general categories. This greatly enabled understanding of the network concept rather than being a barrier to inclusion.

While the questionnaire length and network diagram were changed based on the feedback from the pilot study, all items were retained for further fieldwork. Although items were temporarily deleted during analysis of the pilot data to obtain an initial assessment of validity, these items were retained in the full survey as it was deemed premature for final deletion during analysis of the pilot data. Furthermore, the pilot study was based on a small sample size and conducted in the wine industry.

5.12. Summary

Given the abstract nature of networks and the dearth of past empirical network studies, creativity is essential in order to contribute to theory development in the NM research stream. It is necessary to

93

ensure that the levels of theory, measurement and statistical analysis all return to the network level, although measurement is facilitated via key informants in participating network organizations.

In order to reduce the likelihood of misinterpretation of the frame of reference of networks, screening was conducted and questionnaires administered initially within each network via personal interviews to increase accurate identification of networks. Given the boundary-less nature of networks, issue-based networks and snowballing were adopted. The existing literature was incorporated with findings from the qualitative research in order to develop operational definitions and constructs. This is because of the lack of theory development in NM.

The results from scale purification and evaluation from the pilot study in the wine industry was presented. Support was provided for the reliability and validity of the developed measures. This study offers a methodological contribution to the literature by proposing validated scales to assess power distribution, coordination, harmony, communication and R&D efficiencies of innovation networks. The scales could now be applied to fieldwork in the ICT and biotechnology/ nanotechnology industries as chapter 6 elaborates.

6.0. Chapter Six - Quantitative Results – Fieldwork in the ICT and Biotechnology /Nanotechnology Industries

6.1. Overview

Following scale development and the pilot study in the wine industry, this chapter describes the fieldwork and testing of the model that was carried out in the ICT and Biotechnology/Nanotechnology networks. Prior to testing the hypotheses using data from the ICT and Biotechnology/ Nanotechnology networks via confirmatory factor analysis with structural equation modelling, several steps were taken. First, data was prepared by recoding, dealing with missing data and conducting various checks including those for normality. Second, scales for constructs were purified further at the post-fieldwork stage. Third, assessment of validity and reliability was carried out for all scales. Forth, one factor congeneric models for each construct were tested for fit. Fifth, the structural model was developed using single indicator latent variables and then assessment of overall fit was done. Once fit was established, hypothesis testing followed.

6.2. Structural Equation Modelling

Structural equation modelling (SEM) has become a widely adopted statistical analysis technique in recent times due to its ability to combine confirmatory factor analysis (CFA) and testing of full models through path analysis (Byrne, 2001; Kline, 2005). Consequently, SEM is advantageous as it analyses multiple relationships between independent and dependent variables compared to other multivariate techniques that only examine a single dependence relationship (Hair et al., 2006). It also provides goodness-of-fit tests which assess the level of support given to hypothesized theoretical models (Cunningham, 2008). Additionally, SEM is useful as it represents unobserved concepts and incorporates measurement and structural error in its modelling, thus leading to more accurate estimates (Diamantopoulos, 1994).

Analysis of Moment Structures (AMOS), a software package for analysis of SEM was chosen for use in this study compared to alternatives, such as Linear Structural Relationships (LISREL), because of its ease of use, accessibility, powerful bootstrapping techniques for dealing with non-normal data and popularity given its compatibility with the statistical program SPSS (Arbuckle, 2006; Cunningham, 2008). Hence, SEM and AMOS are used to analyse data from the full survey of this study - the former because of its ability to test full models and provide accurate estimates that incorporate error and the latter due to its wide acceptance and ease of use.

6.3. Data Preparation and Normality Testing

Data preparation was essential in forming a sound basis to the study and was carried out in several steps including recoding, dealing with missing data and testing for normality. Recoding was carried out on all reverse coded items. Missing data was dealt with by using the method of estimation maximization, which introduces less bias than alternative methods, such as mean substitution or regression imputation which may understate variance (Hair et al., 2006).

Normality is a critical assumption underlying the application of SEM, which if violated, leads to an inaccurate assessment of fit, biased parameter estimates and rejection of true models (Anderson and Gerbing, 1988). Normality assumes that variables and their linear combinations are unbiased and consistent (Bollen, 1989). This was checked using tests for skewness and kurtosis (De Carlo, 1997; Tabachnick and Fidell, 2007). Skewness refers to the symmetry of the distributions while kurtosis refers to their peakedness (Hair et al., 2006). Normal distributions have values of kurtosis and skewness of zero. Values above zero indicate that the distribution is relatively peaked while values below zero indicate that the distribution is relatively flat. Acceptable values of skewness and kurtosis should not exceed absolute values of 2 and 7 respectively (West et al., 1995). The constructs exhibited acceptable levels of normality, with skewness and kurtosis ranging from -1.075 to 0.690 and -0.638 to 1.379 as shown in Appendix E. Multivariate normality for both the full models for the ICT and Biotechnology / Nanotechnology samples were less than the upper limit of 7, thus, exhibiting acceptable normality.

Due to the potential effect of non-normality of fit statistics, although normality fell within the acceptable range, a number of techniques were employed to deal with the slight to moderate levels of non-normality. First, recommended fit indices were checked such as the Comparative Fit Index (Lei and Lomax, 2005). Second, the Bollen-Stine bootstrapping method was used when necessary. Bootstrapping is a technique that generates multiple samples from the original sample (Byrne, 2001). This re-sampling is used to re-assess the overall fit of the model, chi-square, and to generate a modified p-Value, the Bollen-Stine p-Value, which is adjusted for lack of multi-variate normality (Bollen and Stine, 1992).

6.4. Scale Purification at Post-fieldwork Stage

Following assessment of normality, scale purification was undertaken at the post-fieldwork stage using confirmatory factor analysis (CFA). While exploratory factor analysis (EFA) was useful in the pilot study, given the novelty of the research to establish the number of dimensions and their related items, CFA was deemed suitable at the post-fieldwork phase (Churchill, 1979). The main reason in choosing CFA at this latter phase is that it is more theoretically driven because only items pertinent to specific dimensions are associated with those factors whereas, all items load on all factors in EFA leading to difficulty in replication of results (Cunningham, 2008; Gorsuch, 1983). All initial items were retained in the full survey rather than deleting them after the pilot study which was based on a small sample size in a specific wine network. Therefore, different items may be more prominent in the other industries investigated in the full survey. Table 18 shows the scale purification post full field work and compares the items used in the pilot study with those used in the analysis of the full survey.

Table 18. Scale Purification Post Full Fieldwork

Construct	Number of Items		Comparison between Pilot and Full Study	
	Pilot	Full Study	# of similar items	# of different items
Power Distribution	3	3	2	1
Coordination	5	6	3	3
Harmony	3	3	3	0
Communication Efficiency	12	8	6	8
R&D Efficiency	6	3	3	3

As anticipated, while there were similar items between the pilot and the full study, there were also several different items used as illustrated in Table 18. Appendices D and E contain the actual items used in these two phases.

6.5. Construct Reliability and Validity at Post-fieldwork Stage

All scales were evaluated for reliability and validity as presented in Table 19.

97

Table 19. Reliability and Validity of Constructs

Construct	Reliability				Convergent Validity	
	Coefficient Alpha		Construct Reliability		Variance Extracted	
	B/N	ICT	B/N	ICT	B/N	ICT
Trust	.944	.953	.874	.874	.501	.500
Power	.746	.832	.747	.750	.500	.501
Coordination	.732	.723	.855	.856	.500	.500
Harmony	.778	.718	.746	.749	.500	.500
Communication Efficiency	.834	.765	.887	.887	.500	.500
R&D Efficiency	.902	.895	.751	.750	.502	.501

Reliability

Reliability was assessed using coefficient alpha and construct reliability. Coefficient alpha was calculated using SPSS 15.0. Although there is little consensus on acceptable levels, values above 0.7 are deemed acceptable (Hair et al., 2006; Kline, 2005). Coefficient alpha for all constructs exceeds this value, thereby demonstrating acceptable reliabilities. Construct reliabilities were calculated using the information from AMOS on standardized item loadings and error measurement from the congeneric models for each construct (Byrne, 2001; Kline, 2005). This information is contained in Appendix F. Convention suggests that construct reliability scores should exceed 0.7 (Hair et al., 2006). All constructs achieved acceptable construct reliability (See Table 19).

Validity

Convergent validity assesses the extent to which measures of the same construct correlate (Churchill, 1979). Using one factor congeneric models, item loadings were acceptable as they all exceeded the threshold of 0.5 (Steenkamp and van Trijp, 1991). Additionally, convergent validity was calculated by examining variance extracted. Data used in the calculation of variance extracted is contained in Appendix F. Evidence of convergent validity was provided for all constructs as variance extracted either equates or exceeds the lower limit of 0.5 (see Table 19).

Assessment of discriminant validity was carried out using a widely used test from Fornell and Larcker (1981) which compares variance extracted of each construct with the square of the highest correlation that each factor shares with other factors (Ramani and Kumar, 2008; Rokkan et al., 2003; Straub, 1989). All factors exhibited discriminant validity as their variances extracted all over 0.500 exceed the square of the highest shared variance between factors which was 0.460.

6.6. Goodness of Fit Indices

Once reliability and validity were assessed, goodness-of-fit indices were applied in order to assess the fit of the measurement and causal models. The Chi-square statistic is the main measure of model fit. It tests the extent to which the data supports the hypothesized model by comparing the sample and model implied matrices of variances and covariances (Cunningham, 2008). A significant chi-square implies that the model does not account for the data whereas a non-significant chi-square (i.e. p-Value >0.05) provides model support.

Although the Chi-square statistic offers a fundamental assessment of model fit, it is sensitive to small sample sizes and non-normality (Kline, 2005). The data collected in this study included adequate sample sizes from each industry of 124 for ICT and 95 for biotechnology/ nanotechnology, as these meet the suggested level of approximately 100 respondents of stable SEM analysis (Hair et al., 2006). Additionally, as discussed in Section 6.3, the model exhibited multivariate normality. Despite sufficient sample size and evidence of normality, the Chi-square statistic was supplemented by other varied tests of fit in order to gain a consensus on the applicability of the model. Table 20 presents the main fit statistics applied and their acceptable levels (Byrne, 2001; Hair et al., 2006; Hu and Bentler, 1998; Kline, 2005).

Table 20. Fit indices used to evaluate fit of structural model

Type of Index	Name of Index	Acceptable Level
Model Fit	Chi-square	
	degrees of freedom (df)	
	P Value	> 0.05
Absolute Fit	Normed Chi-square (Chi-square/df)	< 2
	Goodness-of-Fit (GFI)	>.90
	Adjusted Goodness of Fit (AGFI) (difference between GFI and AGFI)	GFI - AGFI < .6
	Standardized Root Mean-square Residuals (SRMR)	< .05
	Root Mean Square of Approximation (RMSEA)	< .08
Incremental Fit	Comparative Fit Index (CFI)	> .95
	Tucker-Lewis Index (TLI)	> .95

In addition to the Chi-square test for model fit, a range of absolute and incremental fit indices were applied. Absolute fit indices assess the extent to which the specified model reflects the sample data (Cunningham, 2008). The normed Chi-square statistic (i.e. Chi-squared/ degrees of freedom) is a measure of absolute fit and is deemed acceptable if it is less that the acceptable level of 2 and p-values greater than 0.5 (Hair et al., 2006). The Goodness-of-fit Index (GFI) is another well cited absolute fit index in the literature. It indicates the relative degree of covariances of latent variables that are predicted by the model (Mathieu et al., 1992; Schumacker and Lomax, 1996). The AGFI is the GFI adjusted for degrees of freedom and number of variables (Cunningham, 2008). They are widely cited although some opponents to GFI and AGFI argue that they should not be used as they are inconsistent and overly sensitive to sample size. The Root-Mean-Square Error of Approximation (RMSEA) is less sensitive to sample size and distribution and is calculated using the Chi-square statistic, sample size and degrees of freedom (Rigdon, 1996).

Incremental fit indices such as the Comparative Fit Index (CFI) and Tucker-Lewis Index (TLI) compare the improvement in fit of the target model relative to a null model with uncorrelated observed variables (Hu and Bentler, 1998). The TLI has been less sensitive to sample size and more sensitive to misspecification compared to alternative fit indices (Anderson and Gerbing, 1984). Although criticized for its minimal sensitivity to the absence of model fit, the CFI remains a widely reported fit statistic in the literature (Hutchinson, 1998).

6.7. Fit Assessment of Congeneric Models

By applying various measures of model fit, congeneric models for each construct were tested prior to analyzing the causal model. Checking for model fit at the construct level prior to combining them structurally is important for diagnosing and reducing possible amalgamation of problems at that later stage. The model fit for all constructs was acceptable as shown in Table 21. Table 20 discussed in Section 6.6 provides the acceptable levels for all fit statistics. Loadings for all items exceeded 0.50 which is a reflection of convergent validity as shown in Appendix F (Steenkamp and van Trijp, 1991). Only the ICT sample for the construct of communication efficiency required application of bootstrapping to deal with non-normal data. The regular p-Values for all other samples and constructs exceeded the minimum value of 0.05.

100

Table 21. Fit of Congeneric Models

Construct		Chi-sq.	d.f.	Pvalue >.05	GFI >.90	AGFI	GFI-AGFI <.6	CFI >.95	TLI >.95	RMSEA <.08	SRMR <.05
Trust	BN	16.909	14	.261	.962	.923	0.039	.996	.995	.041	.0229
	ICT	20.782	14	.107	.944	.888	0.056	.990	.984	.072	.0200
Power	BN	.743	8	.491	.989	.949	0.04	1.000	1.003	.000	.0318
	ICT	2.509	8	.961	.991	.888	0.103	1.000	.984	.072	.0200
Coord.	BN	6.653	6	.354	.983	.939	0.044	.997	.993	.030	.0287
	ICT	8.078	6	.232	.973	.905	0.068	.992	.980	.061	.0361
Harmony	BN	7.709	8	.462	.980	.948	0.032	1.000	1.003	.000	.0327
	ICT	12.849	8	.117	.957	.891	0.066	.971	.947	.080	.0454
CommEff	BN	18.976	14	.166	.964	.908	0.056	.990	.980	.054	.0252
	ICT	26.655	14	.057	.937	.839	0.098	.966	.932	.098	.0524
R&Deff	BN	.743	8	.491	.980	.949	0.031	1.000	1.003	.000	.0318
	ICT	2.509	8	.961	.991	.978	0.013	1.000	1.034	.000	.0240

6.8. Fit Assessment of Causal Model

After the one factor congeneric models were assessed, a causal model was developed using single indicator latent variables (see Figure 12).

Figure 12. Causal model

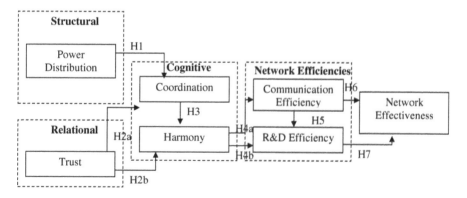

101

Single indicator latent variables were used as a means of data reduction given our small sample size (Rowe, 2002). Furthermore, they were useful as they account for measurement error in the model, and therefore, estimates were less biased compared to the use of composite variables which ignore measurement error (Bollen, 1989). Following Munck (1979), values for the regression coefficient and measurement error were calculated taking into account coefficient alpha (α) and the standard deviation (SD) of the composites using the following formulae:

$$\text{Regression coefficient} = SD \sqrt{\alpha}$$
$$\text{Measurement error variance} = SD^2(1 - \alpha)$$

Appendix G contains correlation matrices and shows that the multicollinearity does not pose a problem as no correlations exceeded the upper limit of 0.9 (Hair et al. 2006). Applying single indicator latent variables, Table 22 shows that the fit achieved of the causal model for each industry was acceptable.

Table 22. Fit of structural model

Type of Index	Name of Index	Acceptable Level	B/N	ICT
Model Fit	Chi-square		16.116	20.576
	degrees of freedom (df)		12	13
	P Value	> 0.05	.186	.082
Absolute Fit	Normed Chi-square (Chi-square/df)	< 2	1.343	1.583
	Goodness-of-Fit (GFI)	>.95	.965	.937
		GFI - AGFI		
	Adjusted Goodness of Fit (AGFI)	< .6	.919	.865
	Standardized Root Mean-square Residuals (SRMR)	< .05	.0221	.0415
	Root-Mean-Square Error of Approximation (RMSEA)	< . 08	.053	.079
Incremental Fit	Comparative Fit Index (CFI)	> .95	.993	.981
	Tucker-Lewis Index (TLI)	> .95	.988	.969

6.9. Hypothesis Tests

Given that the model exhibited good fit, tests of hypotheses emerging from the literature review were deemed suitable. As demonstrated in Table 23, all hypotheses were supported in at least one industry. Patterns emerged as six out the nine hypotheses were supported in both industries.

Table 23. Hypotheses tests

Hypothe sis	Independent Variable	Dependent Variable	B/N Support pValue		ICT Support p Value		Both Industries Supported
H1	Power	Coordination	Supported	.000	Supported	.000	YES
H2a	Trust	Coordination	Supported	.000	Supported	.004	YES
H2b	Trust	Harmony	Supported	.000	Supported	.000	YES
H3	Coordination	Harmony	Supported	.000	Supported	.000	YES
H4a	Harmony	R&D Efficiency	Supported	.000	Not Supported	–	NO
H4b	Harmony	Com Eff	Supported	.000	Supported	.000	YES
H5	Com Eff	R&D Eff	Not Supported	–	Supported	.000	NO
H6	Comm Eff	Network Eff	Supported	.000	Not Supported	–	NO
H7	R&D Eff	Network Eff	Supported	.000	Supported	.000	YES

N.B. Results are based on a 99.9% confidence level

In order to probe more deeply into the type of causal relationships, checks for mediation and partial mediation were carried out using alternative SEM models (Baron and Kenny 1986; Cunningham 2008). Consistent with results in Table 23, all respective variables demonstrated full mediation. For example, coordination mediated the relationship between power distribution and harmony in both industries. Additionally, harmony mediated the relationships between coordination and communication efficiency, and also between the latter and R&D efficiency.

In the ICT industry, communication efficiency mediated the relationship between harmony and R&D efficiency, while R&D efficiency mediated the relationship between communication efficiency and network effectiveness. Alternatively, in the biotechnology/ nanotechnology industry, R&D efficiency mediated the relationship between harmony and network effectiveness, and communication efficiency mediated the relationship between harmony and network effectiveness. No evidence was found for partial mediation.

Figure 13 provides a comparison of significant paths to network effectiveness in the selected industries. The results revealed a number of similarities and a few minor differences across industries on the key factors and their relationships required to effectively manage networks. While the relationships pertaining to the network outcome factors were slightly different, all factors were deemed important in both industries as hypotheses around them were supported.

Figure 13. Comparison of Significant Paths to Network Effectiveness

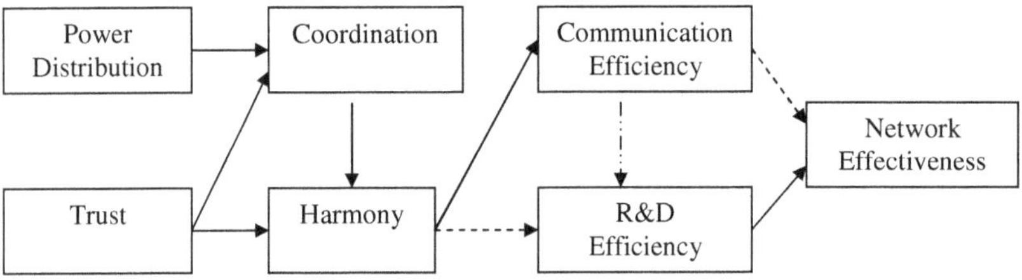

Key
Common Paths ⟶ *ICT only paths* ⟶ *B/N only paths* ⟶

6.9.1. Key factors for NM: Coordination, Harmony and Communication Efficiency

The results in this study confirmed the importance of coordination, harmony and communication efficiency in achieving network outcomes that was suggested in the inter-organizational literature (Denize et al., 2005; Inkpen and Tsang, 2005; Moenaert et al., 2000; Oliver, 1991; Rowley, 1997). This study validates these key constructs and provides empirical evidence for their significance in a network context.

Coordination

The results confirmed the importance of coordination. The study contributes a network perspective to the inter-organizational literature as coordination has traditionally been explored at the relationship level and in particular, in supply chain management research rather than in innovation networks (Axelsson and Easton, 1992; Mohr et al., 1996). Consequently, it provides much needed empirical evidence on network coordination (Medlin, 2006).

104

The findings confirmed that although excessive coordination is unnecessary, some degree of it is required (McCosh et al., 1998; Ojasalo, 2004; Powell, 1990; Williamson, 1991). This finding is similar to earlier empirical results that suggest that a degree of formality is useful between partners in co-marketing alliances (Farrelly and Quester, 2005). The results also validate the need for formalization, adequate coordination and a moderately structured coordinating body. The importance of having a single authority was revealed so that there is ultimate responsibility and accountability in the achievement of outcomes. However, it was not deemed necessary to create a new formal entity as stakeholders may have varied identities and objectives, such as, education, service, profit and research. Thus, the structure should be moderately loose, with collaborating organizations maintaining their identities. This coordinating mechanism should have understanding and representation of all core stakeholders. It should adopt a synchronizing role rather than the traditional management approach based around hierarchies.

Harmony

The findings from this study also confirmed the importance of harmony in an innovation network context. It validated the construct for harmony in a network context as it was previously investigated in an intra-organizational context between the marketing and R&D functions in the exiting literature (Gupta et al., 1986). Furthermore, it contributes a network understanding of its respective dimensions of conflict and cooperation to the marketing literature as these were mainly explored from a dyadic perspective (Welch and Wilkinson, 2005). It therefore, reiterates the need for both conflict and cooperation in the innovation process (Laine, 2002; Vaaland, 2001). It also contributes a more reliable scale as extant scales has been criticised for their low reliabilities (Song and Thieme, 2006).

The findings reveal that collaborations have more successful outcomes when the objectives of the industry [business] partner are included in early stages in setting the research agenda, rather than being solely focused on the objectives of the research [university or research organization] partner. The results also validate the need for open discussion and debate rather than silent disagreement that may not generate the best results. Compromise in decision making was also viewed as necessary in ensuring that each party obtains value from the network.

Communication Efficiency
The results reveal that communication efficiency is an important success factor. Although the network literature contains discussion about the importance of communication efficiency (Ford and Johnsen,

2000; Ford and Johnsen, 2001; Gadde and Hakansson, 2001), sparse empirical evidence has been provided in support of this notion. The study, therefore, contributes a measure of communication efficiency by building on the Moenaert et al. (2000) definition and validating the scale both qualitatively and quantitatively.

The findings confirmed the dimensions suggested by Moenart et al. (2000) which include transparency, codification, credibility and communication costs. The results also provide a construct of acceptable reliability and validity.

6.9.2. Interrelationships among Coordination, Harmony and Communication Efficiency

The study offers strong empirical evidence for the influence of coordination on harmony. The p-Value for the relationship between coordination and harmony was 0.000 from the path analysis, thus, providing consistent support for the resounding impact of coordination and harmony in both the ICT and the biotechnology / nanotechnology networks. These findings confirm the discussion in the literature that coordination may be necessary to ensure that multiple actors could work cohesively (McCosh et al., 1998). Coordination may involve a level of formalization, clear definition of deliverables and a single authority who serves as a network manager. These factors may reduce the likelihood of escalation of conflict to unmanageable levels. Thus, harmony may be maintained.

In turn, the study also provides significant evidence for the positive impact of harmony on communication efficiency. The p-Values for the relationship between harmony and communication efficiency were 0.000 in both the ICT and biotechnology/nanotechnology networks. Harmony involves give-and-take in the relationships with both parties trying to understand the others' view points and incorporate them in early stages when setting the research agenda. Therefore, it is likely that these measures may increase communication efficiency in the network.

Song et al. (2006) establish a link between harmony and the information gap as the latter can be a symptom of a lack of communication efficiency. The information gap is the difference between ideal and achieved levels of information sharing among participants (Song and Thieme, 2006, p 314). This is part of communication efficiency by Moenaert et al. (2000) in the motivation, willingness and ability to share information. Information exchange is an aspect of communication (Denize et al., 2005).

An interesting finding was the mediated relationship between coordination and communication efficiency. The evidence provided in this study indicates that harmony mediates this relationship and no support is offered for the direct relationship between coordination and communication efficiency. This is surprising as the literature suggests that there should be a direct relationship between these constructs although supporting empirical evidence has yet to be provided (Rowley 1997).

6.9.3. Antecedents to Coordination, Harmony and Communication Efficiency

Power Distribution

The study provides evidence for the impact of power distribution on coordination in networks. It first contributes theoretically to the understanding of power in networks by operationalising a network construct for power and validating it both qualitatively and quantitatively. Consequently, this extends the power construct towards a network perspective as it is predominantly explored from a dyadic perspective in the extant inter-organizational literature (Dahl, 1957; Dwyer, 1980; Gaski, 1984; Frazier and Rody, 1991; Hunt and John, 1974; Lusch, 1976; Lusch and Brown, 1982; Welch and Wilkinson, 2005). Based on the quantitative survey, support was offered in both the ICT and biotechnology / nanotechnology networks for the positive impact of power distribution on coordination as reflected in p-Values of 0.000. This reiterates the findings in the literature that coordination improves with density given the increased number of ties (Oliver, 1991).

Trust

In addition to power distribution, the study also provides empirical evidence for the prolific influence of trust on coordination and also for the impact of trust on harmony in networks. In providing a network perspective of trust by incorporating the views of diverse actors, the study contributes and extends understanding towards network level trust. As the extant business and marketing literatures provide predominantly an organizational perspective or that pertaining to one particular type of participant such as CEOs, salespersons or buyers (Medlin and Quester, 2002; Seppanen et al., 2007).

The study validates the construct for trust in a network context by varied players including universities, business and government. The results also provide empirical evidence for the impacts of trust on coordination and harmony in the ICT and biotechnology/nanotechnology industries. This confirms the literature which suggests that trust influences network coordination as it is seen as a network governance mechanism where networks with higher trust levels require less coordination resulting in reduced governance costs (Bidault and Jarillo, 1997; Powell, 1990; Rowley et al., 2000; Seppanen et al., 2007). It also confirms that trust impacts on harmony as it facilitates conflict management as trusting network actors may forego short sighted goals, voice their views openly and focus on developing shared initiatives (Achrol and Kotler, 1999; Powell, 1990; Rowley et al., 2000; Seppanen et al., 2007; Uzzi, 1996).

6.9.4. Network Outcomes

While a number of similarities in success factors between both industries discussed thus far contribute to theory development, differences particularly among outcome variables are also interesting and may provide industry specific implications. As discussed, a number of patterns were uncovered. Evidence was provided for the significant impact of power distribution, trust, coordination, and harmony and communication efficiency on achieving network outcomes in the ICT and biotechnology/ nanotechnology industries.

R&D Efficiency

The results revealed that the role of R&D efficiency and its respective relationships were different between both industries. R&D efficiency appears to have a more pivotal role in the biotechnology/nanotechnology industry compared to the ICT industry. The results from the former industry exhibit a more complex set of relationships reflecting the significant impact of harmony on both R&D efficiency and communication efficiency which both, in turn, impact on network effectiveness. Alternatively, there is a more linear set of relationships among outcome variables in the ICT industry among harmony, communication efficiency, R&D efficiency and network effectiveness.

There may be possible explanations for the differences between these two industries. For instance, it was not completely surprising that R&D efficiency did not play as prominent a role in the Australian ICT industry which involves a high degree of adopting of foreign technology rather than local R&D. This may, consequently, increase the relative significance of communication efficiency given the technology transfer process involved. However, replication of this study in other countries characterised as being more experienced innovators of technology in particular industries may result in significant relationships between harmony and R&D efficiency. While the results are preliminary, given the prominence of innovation networks in many countries and the need to effectively manage them to ensure that public funds are well allocated, a number of important factors are highlighted in this study which should be considered.

Network Effectiveness

The study applies a measure of network effectiveness that incorporates the diverse perspectives of network participants. Extant literature focuses on organizational outcomes rather than network outcomes. Even the general network literature focuses on outcomes to the organization through network involvement and mostly ignores issues of network-level effectiveness (Aldrich and Whetten, 1981; Knoke and Kuklinski, 1982; Marsden, 1990b; Provan and Milward, 1995). Consequently, the use of the Juster scale was ideal in measuring network effectiveness, and particularly, application, in cases of collaborative innovation where the successes of the initiatives are determined by the contribution of several players.

6.10. Summary

This chapter discussed the results of the fieldwork in the ICT and biotechnology / nanotechnology industries within the quantitative phase of research including the process of data preparation and the purification of measures. An assessment of validity and reliability was carried out and the fit of the congeneric and the structural models were evaluated. The results of the hypothesis testing, which was conducted, revealed patterns among key managerial factors in innovation networks between the biotechnology / nanotechnology industry and the ICT industry. A few interesting differences were also uncovered among the outcome variables of efficiency and effectiveness. The following chapter discusses the research contribution in greater detail along with managerial implications and future research directions.

7.0. Chapter Seven – Conclusions, Managerial Implications and Research Directions

7.1. Overview

Over recent decades, the prominence of innovation networks has grown. Despite this increase, few studies have explored key managerial factors operating in these networks. Extant NM literature contains bias towards private sector organizations (Plewa and Quester, 2006). Additionally, both the NM and TT literatures often place emphasis on the firm level (Aurifeille and Medlin, 2007; Provan and Milward, 1995) and ignore the combined perspectives of varied actors focusing instead on only one type of network participant (Bozeman and Gaughan, 2007; Jensen et al., 2007). Consequently, given this limited focus, constructs pertaining to managing from a network perspective are under-developed (Medlin, 2006).

In an attempt to address these shortcomings, this study informs understanding of management of innovation networks. First extant TT and NM literature was critically reviewed in chapters 2 and 3, respectively, for key constructs discussed previously that may be relevant for managing in a network context. Qualitative research was subsequently conducted to determine construct relevance and a causal model for managing innovation networks was presented that incorporated these constructs in chapter 4. The quantitative research design was discussed in chapter 5. Hypotheses were tested and constructs were validated in chapter 6. The findings contribute to theory development as constructs were operationalized and validated both qualitatively and quantitatively incorporating the perspectives of varied network actors.

Furthermore, supporting evidence reflected perspectives from university, business and government, therefore, providing a genuine network focus. This chapter contains a further discussion of the theoretical and methodological contributions of the research. It also provides managerial implications and outlines limitations and future research directions.

7.2. Research Contribution

The present study has made both theoretical and methodological contributions.

7.2.1. Theoretical Contribution

This study contributes to theory development on managing innovation networks in both the NM and TT literatures.

Theoretical Contribution to Technology Transfer Literature

Despite the focus of the TT literature on various factors contributing to successful TT, examination at the network level of analysis remains limited. The extant TT literature has explored individual factors such as culture (Hofstede, 1980; Kedia et al., 2002; Kedia, 1988; Lin and Berg, 2001; Song and Thieme, 2006; Wiley et al., 2006), organizational factors, for example, adaptive ability and knowledge architecture (Gallagher, 2004; Rebentisch and Ferretti, 1995), and relationship factors, such as, organization culture difference and motivation (Gallagher, 2004; Kedia, 1988; Plewa, 2005; Medlin, 2001). However, network factors which bring about communication and R&D efficiencies and effectiveness from the perspective of multiple actors have been under-explored (Auster, 1990; Charles and Howells, 1992; Heikkinen and Tahtinen, 2006; Lin and Berg, 2001; Niosi, 2006; Rebentisch and Ferretti, 1995; Tushman, 2004).

Although empirical evidence of key network factors is limited, the importance of networks has gained recognition in the TT and its related literatures. Emphasis on this phenomenon has been placed in various streams of literature including TT (Auster, 1990; Charles and Howells, 1992; Heikkinen and Tahtinen, 2006; Niosi, 2006; Tushman, 2004), innovation management (Dexter and Nault, 2006; Jelinek and Markham, 2007; Roberts, 2004; Tushman, 2004), triple helix (Etzkowitz and Leydesdorff, 2000; Etzkowitz and Brisolla, 1999; Etzkowitz and Leydesdorff, 1997) and inter-organizational relationships and networks (Hakansson, 1987; Hakansson, 1989; Plewa, 2005).

While its importance has been noted, further research into its management is necessary. Researchers have placed an emphasis on issues at the firm level rather than the combined level of analysis (Provan and Milward, 1995; Medlin, 2006). Others have focused mainly on the views of one type of organization such as universities (Bozeman and Gaughan, 2007; Jensen et al., 2007).

The present study contributes towards a more holistic network understanding of TT in the following ways. It incorporates the perspectives of business, government and university participants, thus, providing a wider network view. It also extends organizational and relational views towards a more panoramic network understanding by including pertinent network level factors, such as, coordination, power distribution, and communication and R&D efficiencies. Therefore, this study contributes to theory development of a network understanding of TT.

Theoretical Contribution to Network Management Literature

The present study contributes theoretically to a net level understanding of NM pertaining to innovation networks. The current debates driving theory development in the NM stream relate to the ontological characteristics that are attributed to networks and the level of analysis employed by researchers.

Traditional researchers from the industrial network approach tend to view networks as boundaryless phenomena (Hakansson and Ford, 2002; Hakansson and Snehota, 1995) that cannot be managed. The NM model by Ford et al. (2002) that emerged from this approach adopts the view that although it may be impossible to manage networks, 'managing in' networks may be possible by coping with, reacting to and managing relationships. This model adopts the perspective of a focal organization operating in a network.

Other researchers including those in strategic management argue that sub-networks, 'nets' with definite boundaries can, in fact, be defined and managed (Brandenburger and Nalebuff, 1996; Brito, 1999; Jarillo, 1993; Moller and Rajala, 2007; Parolini, 1999). Therefore, they focus on analysis at the net level rather than at an organizational perspective. Models emerging from this approach classify nets based on their value proposition and suggest management strategies for each type of network (Moller et al., 2002). This analysis does have limitations as strategies recommended have yet to be linked to network outcomes (Moller and Rajala, 2007).

While both views of the traditional network theorists and the strategic network researchers are indeed insightful in enhancing an understanding of networks, this study is based on the net level of analysis. This is primarily due to the growing importance attributed to assessing whole nets and their effectiveness in both academic and government policy-making quarters. Although the industrial network literature provides rich conceptual multi-layered analysis based on different levels of

112

aggregation of units within the network, the net level perspective that links key managerial factors to net level outcomes remains underdeveloped empirically. The literature generally adopts an organizational perspective based on network involvement with little attention being given to the whole network (Provan and Milward, 1995). Measures, constructs and operational definitions given in the literature remain biased towards organizational antecedents and outcomes rather than reflecting sufficient network level measurement.

As such, this study focused on NM from the perspective of the whole net rather than that of a focal organization. Given the under-exploration of NM at the net level, the literature for relevant constructs was reviewed, qualitative research was conducted to confirm their applicability and measures were developed and validated both qualitatively and quantitatively at this level. The significant results stemming from the study provide support for applicability of key constructs from this perspective including harmony, coordination and power distribution. The tested model also contributed to theory advancement of NM of innovation network from a net level of analysis.

7.2.2. Methodological Contribution

The scarcity of empirical studies on innovation networks is partly due to methodological challenges concerning problems in defining network boundaries, and the subsequent dearth of suitable network scales.

Quantitative Research at the Net Level of Analysis

Empirical studies of networks can be problematic. Previous empirical studies have adopted a limited view of network boundaries of particular types of organizations, ignoring the views held by the variety of network stakeholders (Leseure et al., 2001). Some have primarily adopted a focal organizational perspective and defined the network boundaries in an ego-centric manner that is limited to relationships of one focal organization (Provan and Milward, 1995). Others have only considered mainly the views of one type of participant in the network, such as, university or business participants rather than their combined perspectives (Bozeman et al., 2007; Jensen et al., 2007). However, few studies have incorporated views of different types of network participants. The study contributes a novel methodology that facilitates empirical testing at the net level of analysis by incorporating both qualitative and quantitative research.

The qualitative research includes interviews of key informants, snowballing and triangulation to have an initial understanding of the boundaries of the net. Focusing on a specific network was important for ensuring that a common frame of reference existed between respondents and researchers (Marsden, 1990b). The composition of organizations in this network was determined by interviewing key informants from organizations within the network until a consensus of core players was reached (Perry and Rao, 2007). These findings were also triangulated with information on collaborations from annual reports and public documents from various organizations in the network. Therefore, snowballing was used to identify particular respondents from collaborating organizations (Blaxter et al., 2001). This was deemed a suitable method given the connected nature of networks, in order to more specifically define the boundaries of the population (Sarantakos, 1998). Connectedness is a central feature of networks, and as such, methodologies based on random sampling and independence among units can be deemed inappropriate (Brito, 1999).

The quantitative survey that followed incorporated a network diagram based on the consensus reached in the qualitative phase, multiple respondents from each organization and further snowballing. To maintain the consistent frame of reference, a diagram of this network was included in the research instrument (Kinnear et al., 1996).

Quantitative fieldwork was conducted with the organizations operating in the specified network. Multiple informants from each organization were surveyed to improve the reliability of responses of each organization (Marsden, 1990b). Respondents were asked to identify organizations with which they collaborated, thus, allowing flexibility in identifying new participating organizations with which there are common agreements (Marsden, 1990b; Wasserman and Faust, 1995).

Thus, this novel methodology was instrumental in empirically researching the net level of analysis. Combined qualitative and quantitative methods, snowballing and triangulation with secondary reports were useful in defining the boundaries of the net and in identifying respondents. The network diagram and multiple informants from each organization contributed to the improved reliability of the research.

Pioneering Measures

Once the boundary problem was addressed by identifying the net and its participants, measurement development followed. As networks in previous empirical research were generally defined from an organization perspective (Provan and Milward, 1995) or type of organization (Jensen et al., 2007), existing principles and constructs are skewed towards an organizational perspective. Consequently, the

114

novel methodology employed in this study led to the development of pioneering measures at the net level.

Although measures used to assess whole nets do exist, they remain limited. Measures of innovation nets are biased towards technical and financial rather than social aspects as the extant literature focuses on the former such as R&D tax incentives, patents, R&D expenditure and level of scientific training (Bozeman and Pandey, 1994; Coupe, 2003; Dietz and Bozeman, 2005; Ernst, 1998; Jensen et al., 2007). Jensen et al. (2007) argue that there is a dire need to develop interaction measures of innovation given the widely accepted understanding that interaction within networks fosters innovation.

While the social networks literature does contain measures that assess social dimensions of whole nets, further advancement is necessary. These measures are generally geared towards a structuralist rather than process orientation. Although mathematical and graphical techniques are used, they mainly serve to provide descriptions of network structure with sparse theory development (Cook et al., 1983; Marsden, 1990b; Salancik, 1995). This study builds on some of these measures used to assess power distribution within the network, such as centrality, and extends them for use in hypothesis testing. This is a major first step towards engaging these measures in theory development.

In addition to advancing measures stemming from the social networks literature, this study developed other novel measures. Both qualitative and quantitative research was used to operationalize and validate constructs from a networks perspective by incorporating respondents from university, government and business. A literature review was conducted to obtain an initial set of items for each construct. Interviews and a pre-test were conducted in a range of industries including wine, biotechnology, nanotechnology, ICT/ defence and automotive industries. This assisted in determining the relevance and meaning of constructs and in adjusting and supplementing items from the literature.

A pilot study was then conducted in the wine industry and scale purification was carried out using reliability testing. This was followed by a full survey in the ICT and nanotechnology / biotechnology industries. Confirmatory factor analysis was used to test and validate constructs. Measures of acceptable reliability and validity were developed for power distribution, coordination, harmony, communication efficiency and R&D efficiency. Existing measures were re-validated in a network context for trust.

7.3. Managerial Implications

As the study includes the perspectives of diverse actors, guidelines pertain to a range of actors from the participating sectors. Broad implications may be of interest to organizations concerning key managerial factors necessary for inter-organizational innovation as Table 24 details.

Table 24. Summary of Managerial Implications

Key Factors	Managerial Implications
Foster respect among players and avoid abuses of power	• Players should foster a more balanced power distribution whereby they respect others as they all contribute value to joint network initiatives. • They should refrain from abusing power and using intimidation strategies which may affect underlying relationships.
Implement appropriate levels of coordination	• Although rigid coordination hinders creativity, a moderate degree is required to ensure the goals are achieved. • A single coordinating body is necessary to ensure continuity and the achievement of objectives. • This body should have an understanding or representation of all major collaborators and should adopt a synchronizing, enabling role rather than one of rigid control and bureaucracy.
Encourage harmonious practices	• Industry should be included in early phases when setting the research agenda rather than be purely driven by academia so that standardized outcomes could be assimilated into industry. • During negotiation, meetings or discussions, there should be give-and-take among participants. • Each player should challenge the others if necessary and try to understand the others points of view.
Foster an environment of trust in the networks	• Trust is a critical element in network success. • To develop trust, it is important that players keep promises, exhibit frankness and demonstrate integrity in their dealings.
Improve communication efficiency	• It is important to address issues of transparency, credibility, codification, secrecy and communication costs. • To improve transparency, information should be made available to current or potential collaborators in such a way that patents are not compromised. This information could be provided through relevant modes depending on each case such as a website, information day or via written documents. • These formalized avenues may contribute towards knowledge credibility. Information provided should also be presented in simple language so that it can be assimilated by all collaborators. • To reduce secrecy issues, where appropriate, it is also desirable to have a system for managing intellectual property.
Ensure R&D Efficiency	• Value contribution of each partner should be stressed. • The research revealed that power imbalance in a network may affect its dynamics. Therefore, value contribution should be stressed over political affiliation
Assess network effectiveness	• Assessing network effectiveness is important in ensuring that scarce public funds are well allocated.

Network players should foster a more balanced power distribution whereby they respect others as they all contribute value to joint network initiatives. Additionally, actors should refrain from abusing power and using intimidation strategies which may affect underlying relationships. Furthermore, moderate coordination is crucial. While coordination should not be too rigid and constraining, adequate formalization is necessary. Moderate coordination within networks is also supported by other network researchers (Ojasalo, 2004; Powell, 1990; Williamson, 1991). In particular, a single body is necessary which is accountable for ensuring that innovation objectives in networks are achieved successfully (Charles and Howells, 1992). This body can be an existing, designated organization or a new organization or group consisting of representation from the major network players.

To foster a harmonious environment, business participants should appreciate the necessity of being actively involved and of becoming engaged with their research partners in networks from the beginning. This would allow them not only to be able to initiate projects and become proactive from early phases when research agendas are set, but also to easily assimilate ensuing network outcomes. Adequate negotiation training may also be useful or even fundamental for all network participants for ensuring that healthy give-and-take practices are used among them, opinions are well articulated and promised outcomes are achieved.

Trust is a critical element in network success, and therefore, participants should engage in trustworthy practices, such as, keeping promises, exhibiting frankness and candor and demonstrating integrity. Communication efficiency could be encouraged by addressing transparency via the public availability of information without compromising patents; ensuring credibility via formalized channels in accessible language; and maintaining secrecy by taking adequate steps for managing intellectual property. To promote R&D efficiency, collaborators should be selected based on their value contribution rather than their political affiliation.

Implications may also be useful for organizations involved in inter-organizational innovation including universities, businesses and government agencies. Government agencies may incorporate these findings in R&D grant policies and in designing and managing technology transfer and innovation initiatives across sectors. Given validation from various perspectives, the versatile constructs developed in this study could be employed when comparing the views of network actors and for identifying problems within innovation initiatives. These constructs can also be useful to ascertain discrepancies or agreement in attitudes and in formulating strategies to redress problematic issues such as abuses of

117

power, inappropriate levels of coordination or disharmonious practices. Likewise, university actors, including, technology transfer units responsible for clusters and incubator centres may find these constructs useful and the results of interest in managing similar initiatives. Additionally, businesses operating in innovation networks could also use the findings for extending their understanding of key managerial factors for effective inter-organizational innovation.

7.4. Limitations and Future Research

Interpretation of the result of the study should be undertaken in light of a number of limitations. First, sample sizes of 124 and 95 respectively in ICT and biotechnology / nanotechnology, although adequate for statistical analysis, were relatively small, and thus, larger samples may be useful to strengthen results. While this research offers a viable pioneering methodology to capture net level evidence, future research designs should distinguish and analyse the reliability of responses within organizations and among different types of organizations, such as universities, businesses and government agencies. Sub-group analysis based on the duration of relationships may also be insightful. The sample size in this study does not facilitate this type of sub-group analysis, but future research in these areas will be useful and offer interesting cross-group comparisons.

Second, tests of the proposed constructs and hypotheses in non-university-business-government networks may be insightful. Third, the prominence of innovation networks internationally involved in TT and the need to improve scarce public sector allocations of multi-sectoral research (Provan and Milward, 1995) justify research on an international scale. Furthermore, as innovation infrastructures and their level of development vary across countries (Cohen, 2004), country-specific research is also required to test the applicability of the findings of this study to different national contexts. Finally, one of the major advantages of network research is the ability to incorporate various levels of analysis, including, the organizational, relationship and network levels. This research contributes to the network level of analysis which was previously underdeveloped. However, following previous multi-level studies (Ford et al., 2002; Woodside and Biemans, 2005), future research could combine various levels of analysis and their inter-relationships to provide empirical evidence on success factors for managing inter-organizational innovation.

Nevertheless, this study is a first step towards the validation of key constructs and the examination of the relationships between them from the perspectives of various network actors. Thus, the findings enhance understanding of managing innovation networks from a network perspective.

7.5. Summary

Positioned in a multi-disciplinary context, this research provides a unique theoretical contribution for managing innovation networks from a net perspective to the NM and TT literatures. It employs concepts from these respective literatures as a suitable platform to empirically extend theories beyond organizational biases towards understanding a more panoramic view of the network. Given the novelty of the research it employs qualitative research for developing a conceptual model and for operationalizing measures. Quantitative research is essential for providing the much needed empirical evidence to advance theory on NM. The study reveals key success factors for managing innovation networks including harmony, coordination and communication efficiency, as well as, their antecedents of trust and power distribution.

In addition to the theoretical contribution, the study offers methodological advancements through pioneering net level measures and applying an appropriate methodology to conduct sound quantitative research at the net level. Therefore, it takes a traditionally abstract network concept, reflects on topical debates and addresses the boundary problem. It critically appraises and builds on the extant literature for developing scales and validates these both qualitatively and quantitatively. The pervasive industries of biotechnology / nanotechnology and ICT are investigated to identify common patterns which contribute to strengthening theory while also providing industry specific implications.

Managerial implications are suggested for a range of players involved in innovation networks. Government agencies, research organisations, universities and businesses may all find the findings of this study useful in improving the effectiveness of their collaborations and participation in innovation networks. In particular, this study is valuable from a policy stance for ensuring that scarce R&D funds are well managed. The distinctive theoretical contribution at the net level of analysis along with its corresponding scales and methodology serves as a promising foundation for future research, to offer even further insights on this previously under-explored, yet pertinent perspective given the growing phenomenon of innovation networks.

Bibliography

Achrol, R. S. & Kotler, P. (1999) 'Marketing in the Network Economy'. *Journal of Marketing,* Vol.63 No.4, pp.146-163.

Adams, R., Bessant, J. & Phelps, R. (2006) 'Innovation Management Measurement: A Review'. *International Journal of Management Reviews,* Vol.8 No.1, pp.21-47.

Aldrich, H. & Whetten, D. A. (1981) 'Organizational Sets, Action Sets and Networks: Making the Most of Simplicity'. *Paul C. Nystrom and William H. Starbuck (eds.), Handbook of Organizational Design.* London, Oxford University Press.

Allen, M. P. (1987) 'Economic Interest Groups and the Corporate Elite Structure'. *Social Science Quarterly,* Vol.58 No.4, pp.597-615.

Alvesson, M. & Skoldberg, K. (2000) *Reflexive Methodology: New Vistas for Qualitative Research,* London, Sage Publications.

Anderson, H., Havila, V., Anderson, P. & Halinen, A. (1998) 'Position and Role Conceptualizing Dynamics in Business Networks'. *Scandinavian Journal of Management,* Vol.14 No.3, pp.167-186.

Anderson, J. C. & Gerbing, D. W. (1984) 'The Effect of Sampling Error on Convergence, Improper Solutions, and Goodness-of-Fit Indices for Maximum Likelihood Confirmatory Factor Analysis'. *Psychometrika,* Vol.49 No.2, pp.155-173.

Anderson, J. C. & Gerbing, D. W. (1988) 'Structural Equation Modeling in Practice: A Review and Recommended Two-Step Approach'. *Psychological Bulletin,* Vol.103 No.3, pp.411 - 423.

Araujo, L. & Easton, G. (1996) 'Networks in Socioeconomic Systems'. *Networks in Marketing.* California, London, New Delhi, Sage Publications.

Arbuckle, J. L. (2006) *Amos (Version 7.0) [Computer Program],* Chicago, SPSS.

ARC (2006) www.arc.gov.au, Last Accessed May 31, 2006.

Aulakh, P. S., Kotabe, M. & Sahay, A. (1996) 'Trust and Performance in Cross-Border Marketing Partnerships: A Behavioral Approach'. *International Business Studies,* Vol.27 No.5, pp.1005-1032.

Aurifeille, J. M. & Medlin, C. J. (2007) 'Segmentation for Dyadic Analyses of International Business Relationships'. In Aurifeille, J.-M., Svizzero, S. & Tisdell, C. (Eds.) *Globalization and Partnerships: New Features of Business Alliances and International Co-operation.* London, Nova Science Publishers, Inc.

Auster, E. R. (1990) 'The Interorganizational Environment: Network Theory, Tools and Applications'. *Technology Transfer: A Communication Perspective.* Newbury Park, London, New Delhi, Sage Publications.

Austrade (2007) www.austrade.gov.au, Last Accessed May, 2007.

AWRI (2007) http://www.awri.com.au/infoservice/publications/, Last Accessed November 5, 2007.

Axelsson, B. & Easton, G. (1992) *Industrial Networks: A New View of Reality,* London and New York, Routledge.

Barringer, B. R. & Harrison, J. S. (2000) 'Walking a Tightrope: Creating Value through Interorganizational Relationships'. *Journal of Management,* Vol.26 No.3, pp.367-403.

Baumgartner, H. & Homburg, C. (1996) 'Applications of Structural Equation Modeling in Marketing and Consumer Research: A Review'. *International Journal of Research in Marketing,* Vol.13 No.2, pp.139-161.

Bessant, J., Kaplinsky, R. & Morris, M. (2003) 'Developing Capability Through Learning Networks'. *International Journal of Technology Management & Sustainable Development,* Vol.2 No.1, pp.19-38.

Bidault, F. & Jarillo, C. J. (1997) *Trust in Economic Transactions. In F Bidault P Y. Gomez and G. Marion (Eds), Trust: Firm and Society. Paris: Editions ESKA.*

Biddle, B. J. & Thomas, E. J. (1966) *Role Theory: Concepts and Research,* John Wiley & Sons.

Blankenburg, H. D., Eriksson, K. & Johanson, J. (1999) 'Creating Value Through Mutual Commitment to Business Network Relationships'. *Strategic Management Journal,* Vol.20 No.5, pp.467 - 286.

Blaxter, L., Hughes, C. & Tight, M. (2001) *How to Research,* Buckingham and Philadelphia, Open University Press.

Boje, D. & Whetten, D. A. (1981) 'Effects of Organizational Strategies and Constraints on Centrality and Attributions of Influence in Interorganizational Networks'. *Administrative Science Quarterly,* Vol.26 No.3, pp.378-395.

Bollen, K. A. (1989) *Structural Equations with Latent Variables,* New York, John Wiley & Sons.

Bollen, K. A. & Stine, R. A. (1992) 'Bootstrapping Goodness-of-Fit Measures in Structural Equation Models'. *Sociological Methods in Research,* Vol.21 pp.205-229.

Bonacich, P. (1987) 'Power and Centrality: A Family of Measures'. *The American Journal of Sociology,* Vol.92 No.5, pp.1170-1182.

Borgatti, S. P., Everett, M. & Freeman, L. (1999) http://www.analytictech.com/downloaduc6.htm, Last Accessed November 5, 2007.

Borgström, B. (2005) 'Exploring Efficiency and Effectiveness in the Supply Chain: A Conceptual Analysis'. *Proceedings from the 21st IMP Conference.* Rotterdam, Netherlands.

Bozeman, B. & Gaughan, M. (2007) 'Impacts of Grants and Contracts on Academic Researchers' Interactions with Industry'. *Research Policy,* Vol.36 No.5, pp.694-707.

Bozeman, B., Laredo, P. & Mangematin, V. (2007) 'Understanding the Emergence and Deployment of "Nano" S&T'. *Research Policy,* Vol.36 No.6, pp.807-812.

Bozeman, B. & Pandey, S. (1994) 'Cooperative R&D in Government Laboratories: Comparing the US and Japan'. *Technovation,* Vol.14 No.3, pp.145-160.

Brandenburger, A. M. & Nalebuff, B. J. (1996) *Co-opetition,* New York, Doubleday.

Brass, D. J. (1984) 'Being in the Right Place: A Structural Analysis of Individual Influences in An Organization'. *Administrative Science Quarterly,* Vol.29 No.4, pp.518-539.

Brass, D. J. & Burkhardt, M. E. (1993) 'Potential Power and Power Use: An Investigation of Structure and Behavior'. *Academy of Management Journal,* Vol.36 No.3, pp.441 - 471.

Brito, C. M. (1999) 'Issue-Based Nets: a Methodological Approach to the Sampling Issue in Industrial Networks Research'. *Qualitative Market Research: An International Journal,* Vol.2 No.2, pp.92-102.

Bruner II, G. C., James, K. E. & Hensel, P. J. (2001) *Marketing Scales Handbook: A Compliation of Multi-item Measures,* Chicago, American Marketing Association.

Byrne, B. (2001) *Structural Equation Modeling With AMOS: Basic Concepts, Applications, and Programming,* New Jersey, Lawrence Erlbraum Associates, Inc., Publishers.

Cagliano, R., Caniato, F., Corso, M. & Spina, G. (2002) 'Fostering Collaborative Improvement in Extended Manufacturing Enterprises: a Preliminary Theory'. *Proceedings from the 4th CINet Conference.* Espoo, Finland.

Cameron, K. S. (1986) 'Effectiveness as Paradox: Consensus and Conflict in Conceptions of Organizational Effectiveness'. *Management Science,* Vol.32 No.5, pp.539-553.

Campbell, A. J. & Wilson, D. (1996) 'Managed Networks: Creating Strategic Advantage'. *Networks in Marketing.* California, London, New Delhi, Sage Publications.

Carson, D. & Coviello, N. (1996) 'Qualitative Research Issues at the Marketing/ Entrepreneurship Interface'. *Marketing Intelligence and Planning,* Vol.14 No.6, pp.51-58.

Castells, M. (2000a), *The Information Age: Economy, Society and Culture,* Oxford: Blackwell.

Castells, M. (2000b), 'Materials for an Exploratory Theory of the Network Society', *British Journal of Sociology,* Vol.51 No.1, pp.5-24.

Castells, M. and Himanen, P. (2002), *The Information Society and the Welfare State: The Finnish Model,* Oxford: Oxford University Press.

Chapman, R. L. & Magnusson, M. G. (2006) 'Continuous Innovation, Performance and Knowledge Management: An Introduction'. *Knowledge and Process Management,* Vol.13 No.3, pp.129-

131.

Charles, D. & Howells, J. (1992) *Technology Transfer in Europe - Public and Private Networks*, London, Belhaven Press.

Chesborough, H. (2003) *Open Innovation: The New Imperative for Creating and Profiting from Technology* Boston, Mass, Harvard Business School Press.

Chin, W. W. (1998) 'Partial Least Squares Approach to Structural Equation Modelling.' *Modern Methods for Business Research.* G. A. Marcoulides (Ed.) Lawrence Erlboum Associates, Mahwah, New Jersey.

Chonko, L. B., Howell, R. D. & Bellenger, D. (1986) 'Congruence in Sales Force Evaluations: Relation to Sales Force Perceptions of Conflict and Ambiguity'. *Journal of Personal Selling and Sales Management,* Vol.6 No.1, pp.35-48.

Choudhrie, J., Papazafeiropoulou, A. & Lee, H. (2003) 'A Web of Stakeholders and Strategies: A Case of Broadband Diffusion in South Korea'. *Journal of Information Technology,* Vol.18 No.4, pp.281-290.

Chow, S. & Holden, R. (1997) 'Toward an Understanding of Loyalty: The Moderator Role of Trust'. *Journal of Managerial Issues,* Vol.93 No.3, pp.275-298.

Churchill, G. A. (1979) 'A Paradigm for Developing Better Measures of Marketing Constructs'. *Journal of Marketing Research,* Vol.16 No.1, pp.64-73.

Clark, K. B. & Wheelwright, S. C. (1993) *Managing New Product and Process Development: Text and Cases,* New York, Ontario, The Free Press.

Cohen, G. (2004) *Technology Transfer: Strategic Management in Developing Countries,* New Delhi, Sage.

Cohen, W. M. & Levinthal, D. A. (1990) 'Absorptive Capacity: A New Perspective on Learning and Innovation'. *Administrative Science Quarterly,* Vol.35 No.1, pp.128-152.

Cook, K. S. (1977) 'Exchange and Power in Networks of Inter-Organizational Relations'. *Sociological Quarterly,* Vol.18 No.1, pp.62-81.

Cook, K. S., Emerson, R. M. & Gillmore, M. R. (1983) 'The Distribution of Power in Exchange Networks: Theory and Experimental Results'. *The American Journal of Sociology,* Vol.89 No.2, pp.275-305.

Cooper, D. R. & Schindler, P. S. (2006) *Business Research Methods,* Boston Burr Ridge, IL Dubuque, IA Madison, WI New York, San Francisco, St Louis, Bangkok, Bogota, Kuala Lumpur, Lisbon, London, Madrid, Mexico City, Milan, Montreal, New Delhi, Santiago, Seoul, Singapore, Sydney, Taipei, Toronto, McGraw Hill.

Coote, L., Forrest, E. J. & Tam, T. W. (2003) 'An Investigation into Commitment in Non-Western Industrial Marketing Relationships'. *Industrial Marketing Management,* Vol.32 No.7, pp.595-604.

Corley, E. A., Boardman, P. C. & Bozeman, B. (2006) 'Design and the Management of Multi-institutional Research Collaborations: Theoretical Implications from Two Case Studies'. *Research Policy,* Vol.35 No.7, pp.975-993.

Coupe, T. (2003) 'Science is Golden: Academic R&D and University Patents'. *Journal of Technology Transfer,* Vol.28 No.1, pp.31-46.

Cravens, D. W., Shipp, S. H. & Cravens, K. S. (1994) 'Reforming the Traditional Organization: The Mandate for Developing Networks'. *Business Horizons,* Vol.37 No.4, pp.19-28.

CRC-Polymers (2007) http://www.crcp.com.au/annual_reports.asp, Last Accessed November 5, 2007.

CRC-V(2006)
http://www.crcv.com.au/publications/newsletters/CRCV%20Newsletters/Current%20Newslette rs/4.%20CRCV%20Newsletter%20Jul%20Aug%2006.pdf, Last Accessed December 5, 2007.

Cunningham, E. (2008) *Structural Equation Modeling Using Amos,* Melbourne, Statsline.

Dahl, R. A. (1957) 'The Concept of Power'. *Behavioural Science,* Vol.2 pp.201-218.

Daneels, E. & Kleinschmidt, E. J. (2001) 'Product Innovativeness from the Firm's Perspective: Its Dimensions and Their Relation with Project Selection and Performance'. *Journal of Product Innovation Management,* Vol.18 No.6, pp.357-373.

De Carlo, L. T. (1997) 'On the Meaning and Use of Kurtosis'. *Psychological Methods,* Vol.2 No.3, pp.297-307.

DeBresson, C. & Amesse, F. (2006) 'Networks of Innovators: A Review and Introduction to the Issue'. *Research Policy,* Vol.20 No.5, pp.363-379.

Denize, S., Miller, K. & Young, L. (2005) 'Information Exchange: an Actor, Activity and Resource Perspective'. *Proceedings from the 21st IMP Conference.* Phuket, Thailand.

DEST (2006) www.dest.gov.au, Last Accessed May 31, 2006.

Dexter, A. S. & Nault, B. R. (2006) 'Membership and Incentives in Network Alliances'. *IEEE Transactions on Engineering Management,* Vol.53 No.2, pp.250 - 262.

Di Benedetto, C. A., Calantone, R. J. & Zhang, C. (2003) 'International Technology Transfer - Model and Exploratory Study in the People's Republic of China'. *International Marketing Review,* Vol.20 No.4, pp.446-462.

Diamantopoulos, A. (1994) 'Modelling with LISREL: A Guide for the Uninitiated'. *Journal of Marketing Management,* Vol.10 No.1-3, pp.105-136.

Dietz, J. S. & Bozeman, B. (2005) 'Academic Careers, Patents and Productivity: Industry Experiences as Scientific and Human Capital'. *Research Policy,* Vol.34 No.3, pp.349-367.

Dodgson, M. (1993) 'Learning, Trust and Technological Collaboration'. *Human Relations,* Vol.46 No.1,pp.77-95.

Doney, P. M. & Cannon, J. P. (1997) 'An Examination of the Nature of Trust in Buyer-Seller Relationships'. *Journal of Marketing,* Vol.61 No.2, pp.35-51.

DSTO (2007) http://www.dsto.defence.gov.au/collaboration/, Last Accessed November 5, 2007.

DTI (2007) www.dti.gov.uk, Last Accessed February.

Dwyer, F. R. (1980) 'Channel-Member Satisfaction: Laboratory Insights'. *Journal of Retailing,* Vol.56 No.2, pp.45-65.

Dwyer, F. R., Schurr, P. H. & Oh, S. (1987) 'Developing Buyer-Seller Relationships'. *Journal of Marketing,* Vol.51 No.2, pp.11-27.

Dyer, J. H. & Nobeoka, K. (2000) 'Creating and Managing a High-performance Knowledge-sharing Network: the Toyota case'. *Strategic Management Journal,* Vol.21 pp.345-367.

Ebers, M. & Powell, W. W. (2007) 'Biotechnology: Its Origins, Organization, and Outputs'. *Research Policy,* Vol.36 No.4, pp.433-437.

EIA (2007) http://www.eiaa.asn.au/index.cfm/page/content/contentid/131/menuid/132, Last Accessed November 1, 2007.

Eisenhardt, K. M. (1989) 'Building Theories from Case Study Research'. *Academy of Management Review,* Vol.14 No.4, pp.532 - 551.

Ernst, H. (1998) 'Industrial Research as a Source of Important Patents'. *Research Policy,* Vol.27 No.1, pp.1-15.

Etgar, M. (1978) 'Selection of an Effective Channel Control Mix'. *Journal of Marketing,* Vol.42 pp.53-58.

Etzkowitz, H. & Brisolla, S. N. (1999) 'Failure and Success: The Fate of Industrial Policy in Latin America and South East Asia'. *Research Policy,* Vol.28 No.4, pp.337-445.

Etzkowitz, H. & Leydesdorff, L. (1997) 'Introduction to Special Issue on Science Policy Dimensions of the Triple Helix of University-Industry-Government Relations'. *Science and Public Policy,* Vol.24 No.1, pp.2-5.

Etzkowitz, H. & Leydesdorff, L. (1998) 'The Endless Transition: A "Triple Helix" of University-Industry-Government Relations'. *MINERVA,* Vol.36 No.3, pp.203-208.

Etzkowitz, H. & Leydesdorff, L. (2000) 'The Dynamics of Innovation: from National Systems and "Mode 2" to a Triple Helix of University-Industry-Government Relations'. *Research Policy,* Vol.29 No.2, pp.109-123.

Farrelly, F. & Quester, P. (2005) 'Investigating Large Scale Sponsorship Relationships as Co-Marketing Alliances'. *Business Horizons,* Vol.48 No.1, pp.55-62.

Florida, R. (2002), *The Rise of the Creative Class: And How It's Transforming Work, Leisure, Community and Everyday Life,* New York: Basic Books.

Florida, R. (2005), *The Flight of the Creative Class: The New Global Competition for Talent,* New York: Harper Collins.

Ford, D., Håkansson, H., Snehota, I. & Gadde, L. E. (2002) 'Managing Networks'. *Proceeding of the 18th IMP Conference.* Perth, Australia.

Ford, D. & Johnsen, T. (2000) 'Managing Collaborative Innovation in Complex Networks: Findings from Exploratory Interviews'. *Proceedings from the 16th IMP Conference.* Bath, UK.

Ford, D. & Johnsen, T. (2001) 'Managing Networks of Supplier and Customer Relationships for Technological Innovation: Initial Case Study Findings'. *Proceedings from the 17th IMP Conference.* Oslo, Norway.

Ford, N. M., Walker, J., Orville C & Churchill, J., Gilbert A (1975) 'Expectation-Specific Measures of the Intersender Conflict and Role Ambiguity Experienced by Industrial Salesmen'. *Journal of Business Research,* Vol.3 No.2, pp.95-112.

Fornell, C. & Larcker, D. F. (1981) 'Evaluating Structural Equation Models with Unobservable Variables and Measurement Error'. *Journal of Marketing Research,* Vol.18 No.1, pp.39-50.

Frazier, G. L. & Rody, R. C. (1991) 'The Use of Influence Strategies in Interfirm Relationships in Industrial Product Channels'. *Journal of Marketing,* Vol.55 No.1, pp.52-69.

Freeman, C. (1991) 'Networks of Innovators: A Synthesis of Research Issues'. *Research Policy,* Vol.20 pp.1991.

Freeman, L. C. (1979) 'Centrality in Social Networks: I. Conceptual Clarification'. *Social Networks,* Vol.1 pp.215-239.

Freeman, L. C., Roeder, D. & Mulholland, R. R. (1979) 'Centrality in Social Networks: II Experimental Results'. *Social Networks,* Vol.2 pp.119-141.

Freeman, S. (2001) 'Conflict Management and Exit Strategies In Buyer-Relationships in Foreign Markets : A Case Study of an Australian Citrus Fruit Exporter'. *Proceedings from the 17th IMP conference.* Oslo, Norway, IMP group.

Fritsch, M. (2000) 'Interregional Differences in R&D Activities - An Empirical Investigation'. *European Planning Studies,* Vol.8 No.4, pp.409-427.

Fritsch, M. (2004) 'Cooperation and the Efficiency of Regional R&D Activities'. *Cambridge Journal of Economics,* Vol.28 No.6, pp.829-846.

Fritsch, M. & Meschede, M. (2001) 'Product Innovation, Process Innovation, and Size'. *Review of Industrial Organization,* Vol.19 No.3, pp.335-350.

Furman, J. L., Porter, M. E. & Stern, S. (2002) 'The Determinants of National Innovative Capacity'. *Research Policy,* Vol.31 No.6, pp.899-933.

Gadde, L. E. & Hakansson, H. (2001) *Supply Network Strategies,* Chichester, UK ; New York, Wiley.

Galaskiewicz, J. (1996) 'The "New Network Analysis" and Its Application to Organizational Theory and Behavior'. *Networks in Marketing.* California, London, New Delhi, Sage Publications.

Gallagher, K. S. (2004) 'Limits to Leapfrogging in Energy Technologies? Evidence from the Chinese Automobile Industry'. *Energy Policy,* Vol.34 No.4, pp.384-394.

Ganesan, S. (1994) 'Determinants of Long-Term Orientation in Buyer-Seller Relationships'. *Journal of Marketing,* Vol.58 No.2, pp.1-19.

Gans, J. & Hayes, R. (2004) http://www.ausicom.com/01_cms/details.asp?ID=303, Last Accessed May 8, 2006.

Garcia-Valderrama, T. & Mulero-Mendigorri, E. (2005) 'Content Validation of a Measure of R&D Effectiveness'. *R&D Management,* Vol.35 No.3, pp.311-331.

Garrett, T. C., Buisson, D. H. & Yap, C. M. (2006) 'National Culture and R&D and Marketing Integration Mechanisms in New Product Development: A Cross-Cultural Study Between Singapore and New Zealand'. *Industrial Marketing Management,* Vol.35 No.3, pp.293-307.

Gaski, J. F. (1984) 'The Theory of Power and Conflict in Channels of Distribution'. *Journal of Marketing,* Vol.48 No.3, pp.9-29.

Gatignon, H., Tushman, M. L., Smith, W. & Anderson, P. (2002) 'A Structural Approach to Assessing Innovation: Construct Development of Innovation Locus, Type, and Characteristics'. *Management Science,* Vol.48 No.9, pp.1103-1022.

Ghauri, P. N. & Gronhaug, K. (2005) *Research Methods in Business Studies: A Practical Guide,* Essex, England, Pearson Education Limited.

Gibson, D. V., Williams, F. & Wohlert, K. L. (1990) 'The State of the Field: A Bibliographic View of Technology Transfer'. *Technology Transfer: A Communication Perspective.* Newbury Park, London, New Delhi, Sage Publications.

Golfetto, F., Salle, R., Borghini, S. & Rinallo, D. (2007) 'Opening the Network: Bridging the IMP Tradition and Other Research Perspectives'. *Industrial Marketing Management,* Vol.36 No.7, pp.844-848.

Green, S. G., Gavin, M. B. & Aiman-Smith, L. (1995) 'Assessing a Multidimensional Measure of Radical Technological Innovation'. *IEEE Transactions on Engineering Management,* Vol.42 No.3, pp.203-214.

Guiltinan, J., Ismail, R. & William, R. (1980) 'Factors Influencing Coordination in a Franchise Channel'. *Journal of Research,* Vol.56 No.3, pp.41-58.

Gulati, R. (1999) 'Network Location and Learning: The Influence of Network Resources and Firm Capabilities on Alliance Formation'. *Strategic Management Journal,* Vol.20 No.5, pp.397 - 421.

Gulati, R., Nitin, N. & Akbar, Z. (2000) 'Strategic Networks'. *Strategic Management Journal,* Vol.21 No.3, pp.203-215.

Gupta, A. K., Raj, S. P. & Wilemon, D. (1986) 'A Model for Studying R&D-Marketing Interface in the Product Innovation Process'. *Journal of Marketing,* Vol.50 No.2, pp.7-17.

GWRDC (2007) http://www.gwrdc.com.au/, Last Accessed November 5, 2007.

Hadjikhani, A. & Hakansson, H. (1996) 'Political Actions in Business Networks: A Swedish Case'. *International Journal of Research Marketing,* Vol.13 No.5, pp.431-447.

Hair, J. F., Anderson, R. E., Tatham, R. L. & Black, W. C. (2006) *Multivariate Data Analysis,* New Jersey, Prentice Hall.

Hakansson, H. (1982) *International Marketing and Purchasing of Industrial Goods: An Interaction Approach,* Chichester; New York; Brisbane; Toronto; Singapore, Chichester : Wiley.

Hakansson, H. (1987) *Industrial Technological Development: A Network Approach,* London, Routledge.

Hakansson, H. (1989) *Corporate Technological Behaviour : Co-operation and Networks,* London, New York, Routledge.

Hakansson, H. & Ford, D. (2002) 'How Should Companies Interact in Business Networks?' *Journal of Business Research,* Vol.55 No.2, pp.133-139.

Hakansson, H. & Snehota, I. (1995) *Developing Relationships in Business Networks,* London, Routledge.

Hakansson, H. & Vaaland, T. I. (2000) 'Exploring Interorganizational Conflict In Complex Projects'. *Proceedings from the 16th IMP Conference.* Bath, U.K.

Havila, V. (1992) 'Intermediary and Role Expectations in International Business Relationships'. *Proceedings from the 8th IMP Conference.* Lyon, France.

Hedaa, L. (1999) 'Black Holes in Networks'. *Advances in International Marketing,* Vol.9 pp.131 - 148.

Heikkinen, M. T., Mainela, T., Still, J. & Tahtinen, J. (2006) 'Organizational Roles for Managing in Nets - Case Study of a New Mobile Service Development Net'. *Proceedings from the 22nd IMP Conference.* Milan, Italy.

Heikkinen, M. T., Mainela, T., Still, J. & Tahtinen, J. (2007) 'Roles for Managing in Mobile Service Development Nets'. *Industrial Marketing Management,* Vol.36 No.7, pp.909-925.

Heikkinen, M. T. & Tahtinen, J. (2006) 'Managed Formation Process of R&D Networks'. *International Journal of Innovation Management,* Vol.10 No.3, pp.271-298.

Hofstede, G. (1980) *Culture's Consequences: International Differences in Work Related Values,* Beverly Hills; London; New Delhi, Sage Publications.

Hopkins, M. M., Martin, P. A., Nightingale, P., Kraft, A. & Mahdi, S. (2007) 'The Myth of the Biotech Revolution: An Assessment of Technological, Clinical and Organisational Change'. *Research Policy,* Vol.36 No.4, pp.566-589.

Hu, L. & Bentler, P. M. (1998) 'Fit Indices in Covariance Structure Modeling: Sensitivity to Underparameterized Model Misspecification'. *Psychological Methods,* Vol.3 pp.424-453.

Huhtinen, H. & Virolainen, V. M. (2002) 'Studying Network Management from the Communication Perspective - A Literature Review'. *Proceedings from the 18th IMP Conference.* Dijon, France.

Hunt, S. D. & John, R. N. (1974) 'Power in a Channel of Distribution: Sources and Consequences'. *Journal of Marketing Research,* Vol.11 No.2, pp.186-193.

Iacobucci, D. (1996) *Networks in Marketing,* California, London, New Delhi, Sage Publications.

Inkpen, A. C. & Tsang, E. W. K. (2005) 'Social Capital, Networks and Knowledge Transfer'. *Academy of Management Review,* Vol.30 No.1, pp.146-165.

Jacques, A. (2002) http://millenniumindicators.un.org/unsd/class/intercop/techsubgroup/02-01/tsg0201-8.htm, Last Accessed November, 2006.

Jarillo, J. C. (1993) *Strategic Networks: Creating the Borderless Organization,* Oxford, UK, Butterworth-Heinemann.

Järvelin, A.-M. & Mittilä, T. (2001) 'Expectation Management in Business Relations: Strategies and Tactics'. *Proceedings from the 17th IMP Conference.* Oslo, Norway.

Jelinek, M. & Markham, S. (2007) 'Industry-University IP Relations: Integrating Perspectives and Policy Solutions'. *IEEE Transactions on Engineering Management,* Vol.54 No.2, pp.257 - 267.

Jensen, M. B., Johnson, B., Lorenz, E. & Lundvall, B. A. (2007) 'Forms of Knowledge and Modes of Innovation'. *Research Policy,* Vol.36 No.5, pp.680-693.

132

Jones, C., Hesterly, W. S. & Borgatti, S. P. (1997) 'A General Theory of Network Goevernance: Exchange Conditions and Social Mechanisms'. *Academy of Management Review,* Vol.22 No.4, pp.991-945.

Jung, W. (1980) 'Basic Concepts for the Evaluation of Technology Transfer Benefits'. *Journal of Technology Transfer,* Vol.5 No.1, pp.37-49.

Kafouros, M. (2006) 'The Impact of the Internet on R&D Efficiency: Theory and Evidence'. *Technovation,* Vol.26 No.7, pp.827-835.

Kaltoft, R., Chapman, R., Boer, H., Gertsen, F. & Nielsen, J. S. (2005) 'Collaborative Improvement - Interplay Between the Critical Factors'. *Proceeding of the 6th CINet Conference.* Brighton, United Kingdom.

Kang, S. M. (2007) 'Equicentrality and Network Centralization: A Micro–Macro Linkage'. *Social Networks,* Vol.29 No.4, pp.585-601.

Kedia, B. L., Bhagat, R. S., Harveston, P. D. & Triandis, H. C. (2002) 'Cultual Variations in the Cross-Border Transfer of Organizational Knowledge: An Integrative Framework'. *Academy of Management Journal,* Vol.27 No.2, pp.204-221.

Kedia, B. L. a. B., R. S. (1988) 'Cultural Constraints on Transfer of Technology Across Nations: Implications for Research in International and Comparative Management'. *Academy of Management Review,* Vol.13 No.4, pp.559-571.

Kinnear, T. C., Taylor, J. R., Johnson, L. & Armstrong, R. (1996) *Australian Marketing Research,* New York, San Francisco, Aukland, Bogota, Caracas, Lisbon, London, Madrid, Mexico City, Milan, Montreal, New Delhi, San Juan, Singapore, Tokyo, Toronto, McGraw-Hill Book Company.

Klein, K. J., Dansereau, F. & Hall, R. J. (1994) 'Levels Issues in Theory Development, Data Collection, and Analysis'. *Academy of Management Review,* Vol.19 No.2, pp.195-229.

Kline, R. B. (2005) *Principles and Practice of Structural Equation Modeling,* New York, London, The Guilford Press.

Knight, L. & Harland, C. (2005) 'Managing Supply Networks: Organizational Roles in Network Management'. *European Management Journal,* Vol.23 No.3, pp.281 - 292.

Knoke, D. (1983) 'Organizational Sponsorship and Influence Reputation of Social Influence Associations'. *Social Forces,* Vol.61 No.4, pp.1065-1087.

Knoke, D. & Kuklinski, J. H. (1982) *Network Analysis,* CA, Sage.

Kumar, R. (1996) *Research Methodology: A Step-by-step Guide for Beginners*, Addison Wesley Longman.

Laine, A. (2002) 'Sources of Conflict in Cooperation between Competitors'. *Proceedings from the 18th IMP conference.* Dijon, France.

Laine, A. & Kock, S. (2000) 'A Process Model of Internationalization - New Times Demands New Patterns'. *Proceedings from the 16th IMP Conference.* Bath, UK.

Laursen, K. & Salter, A. (2006) 'Open for Innovation: The Role of Openness in Explaining Innovation Performance Among UK Manufacturing Firms'. *Strategic Management Journal,* Vol.27 No.2, pp.131-150.

Lei, M. & Lomax, R. D. (2005) 'The Effect of Varying Degrees of Nonnormality in Structural Equation Modelling'. *Structural Equation Modelling,* Vol.12 pp.1.

Leseure, M., Shaw, N. & Chapman, G. (2001) 'Performance Measurement in Organisational Networks: An Exploratory Case Study'. *International Journal of Business Performance Management,* Vol.3 No.1, pp.30-46.

Lin, B. & Berg, D. (2001) 'Effects of Cultural Difference on Technology Transfer Projects: An Empirical Study of Taiwanese Manufacturing Companies'. *International Journal of Project Management,* Vol.19 No.5, pp.287-293.

Low, D. & Chapman, R. L. (2003) 'Organisational and National Culture: A Study of Overlap and Interaction in the Literature'. *International Journal of Employment Studies,* Vol.11 No.1, pp.55-75.

Lusch, R. F. (1976) 'Sources of Power: Their Impact on Intrachannel Conflict'. *Journal of Marketing Research,* Vol.13 No.4, pp.382-390.

Lusch, R. F. (1977) 'Franchise Satisfaction: Causes and Consequences'. *International Journal of Physical Distribution,* Vol.7 pp.128-140.

Lusch, R. F. & Brown, J. R. (1982) 'A Modified Model of Power in the Marketing Channel'. *Journal of Marketing Research,* Vol.19 No.3, pp.312-322.

Mani, S. (2002) *Government, Innovation and Technology Policy: An International Comparative Analysis,* Cheltenham, UK; Northampton, MA, USA, Edward Elgar Publishing Limited.

Mariko, S. & Dodgson, M. (2003) 'Strategic Research Partnerships: Empirical Evidence from Asia'. *Technology Analysis & Strategic Management,* Vol.15 No.2, pp.227-246.

Marsden, P. V. (1990a) 'Network Data and Measurement'. *In W. Richard Scott (ed.), Annual Review of Sociology.* Palo Alto, CA, Annual Reviews.

Marsden, P. V. (1990b) 'Network Data and Measurement'. *Annual Review of Sociology,* Vol.16 No.1, pp.435-463.

Mathews, J. A. (2001) 'The Origins and Dynamics of Taiwan's R&D Consortia'. *Research Policy,* Vol.13 No.4, pp.633-652.

Mathieu, J. E., Tannenbaum, S. I. & Salas, E. (1992) 'Influences of Individual and Situational Characteristics on Measures of Training Effectiveness'. *Academy of Management Journal,* Vol.35 No.4, pp.828 - 847.

McAdam, R., Keogh, W., Galbraith, B. & Laurie, D. (2005) 'Defining and Improving Technology Transfer Business and Management Processes in University Innovation Centres'. *Technovation,* Vol.25 No.12, pp.1418-1429.

McCosh, A. M., Smart, A. U., Barrar, P. & Lloyd, A. D. (1998) 'Proven Methods for Innovation Management: An Executive Wish List'. *Journal of Creativity and Innovation Management,* Vol.7 No.4, pp.175 -192.

Medlin, C. J. (2006) 'Self and Collective Interest in Business Relationships'. *Journal of Business Research,* Vol.59 No.7, pp.858-865.

Medlin, C. J. & Quester, P. (2002) 'Inter-firm Trust: Two Theoretical Dimensions Versus a Global Measure'. *Proceedings from the 18th IMP Conference.* Perth, Australia.

Medlin, C. J. & Tornroos, J. A. (2006a) 'Inter-firm Interaction from a Human Perspective'. *Proceedings from the 22nd IMP Conference.* Milan, Italy.

Medlin, C. M. (2001) 'Relational Norms and Relationship Classes: From Independent Actors to Dyadic Interdependence'. *School of Commerce.* Adelaide, University of Adelaide.

Minzberg, H. (1980) *The Nature of Managerial Work,* Prentice-Hall.

Mittilä, T. (2002) 'Whose Expectations Count?' *Proceedings from the 18th IMP conference.* Dijon, France.

Moenaert, R. K., Caeldries, F., Lievens, A. & Wauters, E. (2000) 'Communication Flows in International Product Innovation Teams'. *Journal of Product Innovation Management,* Vol.17 No.5, pp.360-377.

Mohr, J. & Nevin, J. R. (1990) 'Communication Strategies in Marketing Channels: A Theoretical Perspective'. *Journal of Marketing,* Vol.54 No.4, pp.36-51.

Mohr, J. & Ravipreet, S. S. (1995) 'Communication Flows in Distribution Channels: Impact on Assessment of Communication Quality and Satisfaction'. *Journal of Research,* Vol.71 No.4, pp.393-416.

Mohr, J. J., Robert, J. F. & John, R. N. (1996) 'Collaborative Communication in Interfirm Relationships: Moderating Effects of Integration and Control'. *Journal of Marketing,* Vol.60 No.3, pp.103-115.

Moller, K. K. & Halinen, A. (1999) 'Business Relationships and Networks: Managerial Challenge of Network Era'. *Industrial Marketing Management,* Vol.28 No.5, pp.413 - 564.

Moller, K. K. & Rajala, A. (2007) 'Rise of Strategic Nets - New Modes of Value Creation'. *Industrial Marketing Management,* Vol.36 No.7, pp.895-908.

Moller, K. K., Svahn, S., Rajala, A. & Tuominen, M. (2002) 'Network Management as a Set of Dynamic Capabilities'. In group, I. (Ed.) *Proceedings from the 18th IMP Conference.* Dijon, France.

Mollering, G. (2002) 'Perceived Trustworthiness and Inter-firm Governance: Empirical Evidence from the UK Printing Industry'. *Cambridge Journal of Economics,* Vol.26 No.2, pp.139-160.

Monash (2007) http://www.eng.monash.edu.au/mat/CRCP/index.html, Last Accessed November 5, 2007.

Montgomery, J. D. (1998) 'Toward a Role-Theoretic Conception of Embeddedness'. *American Journal of Sociology,* Vol.104 No.1, pp.92-125.

Moon, J. W. & Kim, Y. G. (2001) 'Extending the TAM for a World-Wide-Web Context'. *Information and Management,* Vol.38 No.4, pp.217-231.

Morgan, R. M. & Hunt, S. D. (1994) 'The Commitment-Trust Theory of Relationship Marketing'. *Journal of Marketing,* Vol.58 No.3, pp.20-38.

Mowery, D. C., Nelson, R. R., Sampat, B. N. & Ziedonis, A. A. (2001) 'The Growth of Patenting and Licensing by US Universities: An Assessment of the Effects of the Bayh-Dole Act of 1980'. *Research Policy,* Vol.30 pp.99-120.

Muzzi, C. & Kautz, K. (2004) 'The Diffusion of ICTs in Italian IDs: An Interpretive Study'. *Networked Information Technologies: Difussion and Adoption.* Boston, Kluwer Academic Publishers.

Netemeyer, R. G., Boles, J. S. & McMurrian, R. (1996) 'Development and Validation of Work-Family and Family-Work Conflict Scales'. *Journal of Applied Psychology,* Vol.81 No.4, pp.400-410.

Niosi, J. (2006) 'Introduction to the Symposium: Universities as a Source of Commercial Technology'. *Journal of Technology Transfer,* Vol.31 No.4, pp.399-402.

Nooteboom, B., Berger, H. & Noorderhaven, N. G. (1997) 'Effects of Trust and Governance on Relational Risk'. *Academy of Management Journal,* Vol.40 No.2, pp.308-338.

Norman, P. M. (2002) 'Protecting Knowledge in Strategic Alliances: Resource and Relational Characteristics'. *Journal of High Technology Management Research,* Vol.13 No.2, pp.177-202.

Nunnally, J. C. (1967) *Psychometric Theory,* New York.

Nunnally, J. C. (1970) *Introduction to Psychological Measurement,* New York, McGraw-Hill, Inc.

Nunnally, J. C. (1978) *Psychometric Theory,* New York, McGraw-Hill, Inc.

OECDa (2006) http://www.oecd.org/dataoecd/34/37/2771153.pdf, Last Accessed November, 2006.

OECDb (2006) http://www.oecd.org/document/42/0,2340,en_2649_34409_1933994_1_1_1_1,00.html, Last Accessed November 15, 2006.

Ojasalo, J. (2004) 'Management of Innovation Networks - Two Different Approaches'. *Proceedings from the 20th IMP Conference.* Copenhagen, Denmark.

Oliver, C. (1991) 'Strategic Responses to Institutional Processes'. *Academy of Management Review,* Vol.16 No.1, pp.145-179.

Page, C. & Meyer, D. (2000) *Applied Research Design for Business and Management,* Sydney, McGraw-Hill Companies, Inc.

Park, S. H. & Ungson, G. R. (1997) 'The Effect of Partner Nationality, Organizational Dissimilarity, and Economic Motivation on the Dissolution of Joint Ventures.' *Academy of Management Journal,* Vol.39 No.2, pp.279-309.

Parolini, C. (1999) *The Value Net: A Tool for Competitive Strategy,* Chichester, UK, John Wiley & Sons Ltd.

Perry, C. & Rao, S. (2007) *Convergent Interviewing: A Starting Methodology for an Enterprise Research Program, Innovative Methodologies in Enterprise Research, Hine, D. and Carson, D.,* Northampton, Edward Elgar Publishing.

Pisano, G. (2006) *Science Business: The Promise, The Reality and the Future of Biotech,* Boston, Harvard Business Review.

Plank, R. E., Reid, D. A. & Pullins, E. B. (1999) 'Perceived Trust in Business-to-Business Sales: A New Measure'. *Journal of Personal Selling and Sales Management,* Vol.19 No.3, pp.61-71.

Plewa, C. (2005) 'Key Drivers of University-industry Relationships and the Impact of Organizational Culture Difference; A Dyadic Study'. *School of Commerce.* Adelaide, University of Adelaide.

Plewa, C. & Quester, P. (2006) 'The Effect of a University's Market Orientation on the Industry Partner's Relationship Perception and Satisfaction'. *International Journal on Technology Intelligence and Planning,* Vol.2 No.2, pp.160-177.

Porter, M. E. (1974) 'Consumer Behavior, Retailer Power, and Market Performance in Consumer Goods Industries'. *The Review of Economics and Statistics,* Vol.56 No.4, pp.419-436.

Porter, M. E. (1987) 'From Competitive Advantage to Corporate Strategy'. *Harvard Business Review,* Vol.65 No.3, pp.43-59.

Powell, W. W. (1990) 'Neither Market nor Hierarchy: Network Forms of Organization'. *Research in Organizational Behavior,* Vol.12 pp.295-336.

Powell, W. W., Koput, K. W. & Smith-Doerr, L. (1996) 'Interorganizational Collaboration and the Locus of Innovation: Networks of Learning in Biotechnology'. *Administrative Science Quarterly,* Vol.41 No.1, pp.116-145.

Provan, K. G. & Milward, H. B. (1995) 'A Preliminary Theory of Interorganizational Network Effectiveness: A Comparative Study of Four Community Mental Health Systems'. *Administrative Science Quarterly,* Vol.40 No.1, pp.1-33.

Provan, K. G. & Milward, H. B. (2001) 'Do Networks Really Work? A Framework for Evaluating Public Sector Organizational Networks'. *Public Administration Review,* Vol.61 No.4, pp.414-424.

Ramani, G. & Kumar, V. (2008) 'Interaction Orientation and Firm Performance'. *Journal of Marketing,* Vol.72 No.1, pp.27-45.

Ramirez, R. (1999) 'Value Co-production: Interllectual Origins and Implications for Practice and Research'. *Strategic Management Journal,* Vol.20 No.1, pp.49 - 66.

Rampersad, G., Quester, P., & Troshani, I. (In-Press) "Managing Innovation Networks: A Cross-Industry Investigation of ICT and Nano-Bioscience Networks", Industrial Marketing

Management, Available online http://dx.doi.org/10.1016/j.indmarman.2009.07.002 Last accessed October 2009

Rampersad, G., Quester, P., & Troshani, I. (2009) "Management of Networks involving Technology Transfer from Public to Private Sector: A Conceptual Framework", International Journal of Technology Transfer and Commercialisation, Vol. 8, No. 2/3, pp. 121-141.

Rebentisch, E. S. & Ferretti, M. (1995) 'A Knowledge Asset-based View of Technology Transfer in International Joint Ventures'. *Journal of Engineering and Technology Management,* Vol.12 No.1-2, pp.1-25.

Richards, L. (2002) *Using N6 in Qualitative Research,* Australia, QSR International Pty Ltd.

Ritter, T. & Gemunden, H. G. (2003) 'Network Competence: Its Impact on Innovation Success and Its Antecedents'. *Journal of Business Research,* Vol.56 No.9, pp.745-755.

Ritter, T., Wilkinson, I. F. & Johnston, W. I. (2004) 'Managing in Complex Business Networks'. *Industrial Marketing Management,* Vol.33 No.3, pp.175-183.

Rizzo, J. R., House, R. J. & Lirtzman, S. I. (1970) 'Role Conflict and Ambiguity in Complex Organizations'. *Administrative Science Quarterly,* Vol.15 No.2, pp.150-164.

Roberts, E. B. (2004) 'A Perspective on 50 Years of the Engineering Management Field'. *IEEE Transactions on Engineering Management,* Vol.51 No.4, pp.398 - 403.

Robinson, D. K. R., Rip, A. & Mangematin, V. (2007) 'Technological Agglomeration and the Emergence of Clusters and Networks in Nanotechnology'. *Research Policy,* Vol.36 No.6, pp.871-879.

Rogers, E. M. (1962) *Diffusion of Innovations*, Free Press, New York.

Roijakkers, N. & Hagedoorn, J. (2006) 'Inter-firm R&D Partnering in Pharmaceutical Biotechnology since 1975: Trends, patterns, and Networks'. *Research Policy,* Vol.35 No.3, pp.431-446.

Rokkan, A. I., Heide, J. B. & Wathne, K. (2003) 'Specific Investments in Marketing Relationships'. *Journal of Marketing Research,* Vol.40 No.2, pp.210-224.

Rosenberg, N. & Stern, L. W. (1974) 'A New Approach to Distribution Conflict Management'. *Business Horizons,* Vol.8 pp.437-442.

Rothaermel, F. T. & Thursby, M. (2007) 'The Nanotech Versus the Biotech Revolution: Sources of Productivity in Incumbent Firm Research'. *Research Policy,* Vol.36 No.6, pp.832-849.

Rowe, K.(2002)
http://www.acer.edu.au/research/programs/documents/MeasurementofCompositeVariables.pdf, Last Accessed October, 2006.

Rowley, T., Behrens, D. & Krackhardt, D. (2000) 'Redundant Governance Structures: an Analysis of Structural and Relational Embeddedness in the Steel and Semiconductor Industries'. *Strategic Management Journal,* Vol.21 No.3, pp.369-387.

Rowley, T. J. (1997) 'Moving Beyond Dyadic Ties: A Network Theory of Stakeholder Influences'. *Academy of Management Review,* Vol.22 No.4, pp.887-910.

Ruekert, R. W. & Walker, J., Orville C (1987) 'Marketing's Interaction with Other Functional Units: A Conceptual Framework and Empirical Evidence'. *Journal of Marketing,* Vol.51 No.1, pp.1-19.

Ruttan, V. W. (2001) *Technology, Growth and Development: An Induced Innovation Perspective,* New York, Oxford, Oxford University Press.

Ryan, M. H. (2004) 'The Role of National Culture in the Space Based Technology Transfer Process'. *Comparative Technology Transfer and Society,* Vol.2 No.1, pp.31-66.

Sako, M. & Helper, S. (1998) 'Determinants of Trust in Supplier Relations: Evidence from the Automotive Industry in Japan and in the United States'. *Journal of Economic Behaviour and Organization,* Vol.34 No.4, pp.387-417.

Salancik, G. R. (1995) 'WANTED: A Good Network Theory of Organization'. *Administrative Science Quarterly,* Vol.40 No.2, pp.345-349.

Salmi, A., Anderson, H., Andersson, P. & Havila, V. (2000) 'Business Network Dynamics and M&As'. *Proceeding from the 16th IMP Conference.* Bath, UK.

Sarantakos, S. (1998) *Social Research,* Australia, Macmillan Education.

Schumacker, R. E. & Lomax, R. G. (1996) *A Beginner's Guide to Structural Equation Modeling*, Mahwah, NJ, Erlbaum.

Seppanen, R., Blomqvist, K. & Sundqvist, S. (2007) 'Measuring Inter-Organizational Trust - A Critical Review of the Empirical Research in 1990-2003'. *Industrial Marketing Management*, Vol.36 No.2, pp.249-265.

Shenkar, O., Ronen, S., Shefy, E. & Han-sui Chow, I. (2004) 'The Role Structure of Chinese Managers'. *Human Relations*, Vol.51 No.1, pp.51-72.

Singh, J. & Rhoads, G. K. (1991) 'Boundary Role Ambiguity in Marketing-Oriented Positions: A Multidimensional, Multifaceted Operationalization'. *Journal of Marketing Research*, Vol.28 No.3, pp.328-338.

Smith, J. B. & Barclay, D. W. (1997) 'The Effects of Organizational Differences and Trust on the Effectiveness of Selling Partner Relationships'. *Journal of Marketing*, Vol.61 No.1, pp.3-21.

Snijders, T. A. B. (1981) 'The Degree Variance: An Index of Graph Heterogeneity'. *Social Networks*, Vol.3 pp.163-174.

Snow, C. C. & Miles, R. E. (1992) 'Managing 21st Century Network Organization'. *Organizational Dynamics*, Vol.20 No.Winter, pp.5-20.

Song, M. & Thieme, R. J. (2006) 'A Cross-National Investigation of the R&D-Marketing Interface in the Product Innovation Process'. *Industrial Marketing Management*, Vol.35 No.3, pp.308-322.

Sonquist, J. A. & Koenig, T. (1975) 'Interlocking Directorates in the Top U.S. Corporations: A Graph Theory Approach'. *Insurgent Sociologist*, Vol.5 pp.196-229.

Soosay, C. A. & Chapman, R. L. (2006) 'An Empirical Examination of Performance Measurement for Managing Continuous Innovation in Logistics'. *Knowledge and Process Management*, Vol.13 No.3, pp.192-205.

Spann, M. S., Adams, M. & Sounder, E. (1995) 'Measures of Technology Transfer Effectiveness: Key Dimensions and Differences in their use by Sponsors, Developers and Adopters'. *IEEE Transactions on Engineering Management*, Vol.42 No.1995, pp.19-29.

Stacey, R. (1996) *Complexity and Creativity in Organizations*, San Francisco, CA, Berret-Koehler Publishers.

Steenkamp, J. B. E. M. & Baumgartner, H. (2000) 'On the Use of Structural Equation Models for Marketing Modeling'. *International Journal of Research in Marketing*, Vol.17 No.2/3, pp.195-102.

Steenkamp, J. B. E. M. & van Trijp, H. C. M. (1991) 'The Use of LISREL in Validating Marketing Constructs'. *International Journal of Research in Marketing*, Vol.8 No.4, pp.283-299.

Straub, D. W. (1989) 'Validating Instruments in MIS Research'. *MIS Quarterly*, Vol.13 No.2, pp.147-170.

Sutton-Brady, C. (2000) 'Towards Developing a Construct of Relationship Atmosphere'. *Proceedings from the 16th IMP Conference*. Bath, U.K.

Swan, J., Goussevskaia, A., Newell, S., Robertson, M., Bresnen, M. & Obembe, A. (2007) 'Modes of Organizing Biomedical Innovation in the UK and US and the Role of Integrative and Relational Capabilities'. *Research Policy*, Vol.36 No.4, pp.529-547.

Swann, J. P. (1988) *Academic Scientists and the Pharmaceutical Industry*, Baltimore, John Hopkins University Press.

Sydow, J. & Windeler, A. (1998) 'Organizing and Evaluating Interfirm Networks: A Structurationist Perspective on Network Processes and Effectiveness'. *Organization Science*, Vol.9 No.3, pp.265-285.

Tabachnick, B. G. & Fidell, L. S. (2007) *Using Mutlivariate Statistics*, Massachusettes, Allyn and Bacon.

Ticehurst, C. G. & Veal, T. R. (2000) *Business Research Methods: A Managerial Approach*, Australia, Pearson Education Pty Limited.

Tushman, M. L. (2004) 'From Engineering Management/R&D Management, to the Management of Innovation, to Exploiting and Exploring Over Value Nets: 50 Years of Research Initiated by the IEEE-TEM'. *IEEE Transactions on Engineering Management*, Vol.51 No.4, pp.409-411.

UNCTAD (2005) 'United Nations Conference of Trade and Development, World Investment Report: Transnational Corporations and the Internationalization of R&D'. New York and Geneva, United Nations.

Uzzi, B. (1996) 'The Sources and Consequences of Embeddedness for the Economic Performance of Organizations: The Network Effect'. *American Sociological Review,* Vol.61 No.4, pp.674-698.

Vaaland, T. I. (2001) 'Conflict in Business Relations. The Core of Conflict in Oil Industrial Development Projects.' *Proceedings from the 17th IMP Conference.* Oslo, Norway.

Van de Ven, A. & Walker, G. (1984) 'The Dynamics of Interorganizational Coordination'. *Administrative Science Quarterly,* Vol.29 No.4, pp.598-621.

Van de Ven, A. H. (1976) 'On the Nature, Formation, and Maintenance of Relations Among Organizations'. *Academy of Management Review,* Vol.4 pp.24-36.

Vilkinas, T. & Cartan, G. (2001) 'The Behavioural Control Room for Managers: The Integrator Role'. *Leadership and Organizational Development Journal,* Vol.22 No.4, pp.175-185.

Wasserman, S. & Faust, K. (1995) *Social Network Analysis: Methods and Applications,* Cambridge, New York USA, Melbourne Australia., Cambridge University Press.

Welch, C. & Wilkinson, I. (2005) 'Network Perspectives on Interfirm Conflict: Reassessing a Critical Case in International Business'. *Journal of Business Research,* Vol.58 pp.205-213.

West, S. G., Finch, J. F. & Curran, P. J. (1995) 'Structural Equation Models with Nonnormal Variables'. *R.H. Hoyle (Ed.), Structural Equation Modeling: Concepts, Issues and Applications.* Newbury Park, CA, Sage.

White, H., Boorman, S. & Breiger, R. (1976) 'Social Structure from Multiple Networks - Blockmodels of Roles and Positions'. *American Journal of Sociology,* Vol.81 pp.730-779.

Wiley, J., Wilkinson, I. & Young, L. (2006) 'The Nature, Role and Impact of Connected Relations: A Comparison of European and Chinese Suppliers' Perspectives'. *Journal of Business and Industrial Marketing,* Vol.21 No.1, pp.3-13.

Wilkinson, I., Ritter, T. & Johnston, W. (2004) 'Ability to Manage in Business Networks: A Review of Concepts'. *Industrial Marketing Management,* Vol.33 No.3, pp.175-183.

Williams, F. & Gibson, D. V. (1990) *Technology Transfer: A Communication Perspective,* Newbury Park, London, New Delhi, Sage Publications.

Williamson, O. E. (1975) *Markets and Hierarchies,* New York, The Free Press.

Williamson, O. E. (1991) 'Comparative Economic Organization: The Analysis of Discrete Structural Alternatives'. *Administrative Science Quarterly,* Vol.36 No.2, pp.269-296.

Woodside, A. G. & Biemans, W. G. (2005) 'Modeling Innovation, Manufacturing, Diffusion and Adoption/ Rejection Processes'. *Journal of Business and Industrial Marketing,* Vol.20 No.7, pp.380 - 393.

Xie, J., Song, M. & Stringfellow, A. (1998) 'Interfunctional Conflict, Conflict Resolution Styles, and New Product Success: A Four Culture...' *Management Science,* Vol.44 No.12, pp.192 - 207.

Yi, M. Y., Jackson, J. D., Park, J. S. & Probst, J. C. (2006) 'Understanding Information Technology Acceptance by Individual Professionals: Toward and Integrative View'. *Information and Management,* Vol.43 No.3, pp.350-363.

Yin, R. K. (2003) *Case Study Research: Design and Methods,* Thousand Oaks, London, New Delhi, Sage Publications.

Young-Ybarra, C. & Wiersema, M. F. (1999) 'Strategic Flexibility in Information Technology Alliances: The Influence of Transaction Cost Economics and Social Exchange Theory'. *Organization Science,* Vol.10 No.4, pp.439-459.

Zaheer, A., McEvily, B. & Perrone, V. (1998) 'Does Trust Matter? Exploring the Effects of Interorganizational and Interpersonal Trust on Performance'. *Organization Science,* Vol.9 No.2, pp.141-159.

Zhang, A., Zhang, Y. & Zhao, R. (2003) 'A Study of the R&D Efficiency and Productivity of Chinese Firms'. *Journal of Comparative Economics,* Vol.31 No.3, pp.444-464.

Zolkiewski, J. (2001) 'The Complexity of Power Relationships within a Network'. *Proceedings from the 17th IMP Conference.* Oslo, Norway.

Zurcher, L. A. (1983) *Social Roles: Conformity, Conflict and Creativity,* London, Sage Publications.

Appendices

Appendix A: Information Sheet for Interview

<Name><Title><Address>

<Date>

Dear <Name>,

SCHOOL OF COMMERCE

Security House
233 North Terrace
THE UNIVERSITY OF ADELAIDE
SA 5005
AUSTRALIA

RE: Research on Managing collaborative innovation networks

I am a PhD student from the University of Adelaide. I am currently conducting research on collaborative innovation networks. The results of this research will be part of my PhD thesis. Collaborative innovation networks consist of relationships geared towards creating or transferring technological innovations. They vary in formality and may comprise of combinations of organisations including firms, research institutions and government agencies. The research to be carried out has the following objectives:

Objectives of the research
1. To identify the key factors in managing collaborative innovation networks
2. To determine measures of efficiency and success in managing these networks

Activities to be carried out
In terms of methodology, I am focusing on a few industries – ICT being one of my main case studies. I will then explore the network dynamics within each industry. In order to do this, I will conduct interviews with representatives of the key organisations that collaborate on innovation and R&D within these industries. The interviews will also allow me to construct maps that illustrate key players and relationships within these networks. Once I have an understanding of the key actors in the networks, surveys will then be conducted on the factors for effective management and evaluation of these networks.

Benefits of the research
This research is significant because of the possible benefits it can provide to practitioners and policy makers in increasing the effectiveness of innovation collaborations and technology transfer. Upon project completion, the contributors will be provided with a report of the implications from collected data.

It would be extremely useful if I could meet with you to gain a better understanding of the key actors and relationships in the ICT industry. Any information that you provide will be treated confidentially and will not be associated with yourself in any subsequent publication unless this is expressly requested. If you require any further information, please do not hesitate in contacting me or my supervisor at the contact details listed at the bottom of this page. I have also included an 'Independent Complaints Form' that you could use if you have any concerns about this research.

I look forward to your response. Thank you very much for your time and consideration.

Yours faithfully,
Giselle Rampersad

GISELLE RAMPERSAD
Tel : +61 8 8303 8249 (w) 61 4 0388 6324 (m)
Fax : +61 8 8303 4368
e-mail: giselle.rampersad@adelaide.edu.au

DR INDRIT TROSHANI
Ph   : +61 8 8303 5526
Fax : +61 8 8303 4368
e-mail: indrit..troshani@adelaide.edu.au

Appendix B: Interview Protocol Themes

Interview Clusters	Individual Criteria
Identification of Networks	Definition of Networks • Inter-organizational networks that may include members of government, university and businesses. • Ties that range in formality from subcontracting to informal relationships • Continuous rather than one-off projects • Collaborate to facilitate innovation and TT Identification of TT networks • In industry • Specific to organization
Success factors in Network Management	Based on experience
Performance criteria	Criteria used in practice to judge effectiveness of network
Conceptual Framework	Assessment of Relevance of factors • Power • Harmony • Coordination • Role Expectation • Transfer Scope • Communication Efficiency • R&D Efficiency • Others

Appendix C Questionnaire for Quantitative Pilot Study

SCHOOL OF COMMERCE

Security House
233 North Terrace
THE UNIVERSITY OF ADELAIDE
SA 5005
AUSTRALIA

<Name>
<Title>
<Address>

<Date>

Dear <Name>,

Dear Sir or Madam:

RE: Research on Managing collaborative innovation networks

I am a PhD student from Adelaide University. I am currently conducting a survey on **collaborative innovation networks involving technology transfer** as the central part of my PhD thesis. If you are involved in technology transfer and innovation and you belong to a business, research organization or government agency, your participation in this survey will be greatly appreciated. The research to be carried out aims to identify the key factors in managing collaborative innovation networks.

This research is significant because of the possible benefits it can provide to practitioners and policy makers in increasing the effectiveness of innovation collaborations and technology transfer. Upon project completion, the contributors will be provided with a report that summarises the implications from collected data. All that is required is for you to complete the enclosed questionnaire. It should only take 15 -20 minutes of your time.

Please be assured that the information you provide will remain strictly confidential. It is very important that you answer all questions, even if some appear similar, to ensure that your questionnaire can be included in the research. After completing the questionnaire, please use the prepaid envelope to return it to me, Giselle Rampersad, preferably before March, 2007. If you require any further information, please do not hesitate in contacting me or my supervisor at the contact details listed at the bottom of this page. I have also included an 'Independent Complaints Form' that you could use if you have any concerns about this research.

Clarification of terms used in this questionnaire:

Network: Group of organizations that come together for innovation and technology transfer purposes. It may comprise of ties that vary in formality from contracting to informal relationships. It may come together on a continuous basis rather than only for one-off projects. The terms network, collaboration and collaborative network are used interchangeably.

Participant: An organization belonging to the network. It may be a government agency, research organization or business.

ICT: Information and Communications Technology.

Thank you very much for your time and valuable contribution to this research.

Yours faithfully,
Giselle Rampersad
MSc Durham University, UK

GISELLE RAMPERSAD
Tel : +61 8 8303 8249 (w) 61 4 0388 6324 (m)
Fax : +61 8 8303 4368
e-mail: giselle.rampersad@adelaide.edu.au

DR INDRIT TROSHANI
Ph : +61 8 8303 5526
Fax : +61 8 8303 4368
e-mail: indrit .troshani@adelaide.edu.au

Survey Questionnaire

Details

Your Details
Please fill in the following details.

Name

Organization

Please place an X in **one or more** of the squares to indicate your answer to each question.

Organization Type: ☐ Business ☐ Government ☐ University ☐ Research Organization

Industry Sector: ☐ Biotechnology ☐ ICT ☐ Wine ☐ Other

Location of your office:

☐ Australian Capital Territory ☐ Queensland ☐ Victoria

☐ New South Wales ☐ Tasmania ☐ Northern Territory

☐ South Australia

Network Identification

Questions relate to ONE innovation network in which you in the capacity of your role on behalf of your organization were involved. You may be involved in several different networks. The focus here should be on one network where participants interacted with each other. Kindly note that this network should have involved innovation and technology transfer. It may have comprised businesses, universities, research organizations and government agencies. You may have had relationships with participants that vary in formality from formal contracting to informal relationships. They may have been in existence on a continuous basis rather than only one-off projects. The following network reflects a portion of a network that your organization may have been involved in.

AWRI – Australian Wine Research Institute
CRCs – Cooperative Research Centres (e.g. CRC for Irrigation Futures or CRC-Viticulture)
CSIRO – Commonwealth Scientific and Industrial Research Organization
GWRDC – Grape and Wine Research Development Corporation
SARDI – South Australian Research and Development Institute

Characteristics of Technology Transferred

The following questions relate to the nature of the technology transferred within this network. Keeping in mind the particular network, please indicate to what extent you agree or disagree with the following statements. Please circle only **one response per statement.**

			Strongly Disagree					Strongly Agree
A1	The innovation resulted in a new/ enhanced product.	n/a	1	2	3	4	5	6	7
A2	The innovation improved a service.	n/a	1	2	3	4	5	6	7
A3	The innovation brought about changes in processes.	n/a	1	2	3	4	5	6	7
A4	The innovation resulted in incremental improvements to existing technologies.	n/a	1	2	3	4	5	6	7
A5	The innovation was radically new compared to existing technologies.	n/a	1	2	3	4	5	6	7

Network Processes

The following statements relate to processes of the network such as the distribution of power, coordination, harmony, communication and R&D efficiencies. Please circle only **one response per statement.**

			Strongly Disagree					Strongly Agree
Power Distribution									
B1	One or more large participants dominated the network.	n/a	1	2	3	4	5	6	7
B2	The power distribution in the network was even.	n/a	1	2	3	4	5	6	7
B3	My organization had the same amount of power as the other participants' organizations.	n/a	1	2	3	4	5	6	7
B4	A particular participant had tremendous influence over the other players in the network.	n/a	1	2	3	4	5	6	7
B5	A particular participant could have invoked political or media action.	n/a	1	2	3	4	5	6	7
B6	A particular participant had more control than others over resources such as funding, equipment, skills and competencies.	n/a	1	2	3	4	5	6	7
Harmony									
C1	During negotiation, meetings or discussions, there was give-and-take among participants. Each challenged the others and tried to understand the others points of view.	n/a	1	2	3	4	5	6	7
C2	The research institution and the industry partner were involved in the early phases of discussion in setting the research agenda.	n/a	1	2	3	4	5	6	7
C3	Conflicts between participants were resolved locally among the disagreeing participants rather than via escalation throughout the wider network.	n/a	1	2	3	4	5	6	7
C4	There was compromise among participants in decision-making and each party obtained value from the network.	n/a	1	2	3	4	5	6	7
C5	In the event of tensions or disagreement, an effective conflict resolution mechanism was in place.	n/a	1	2	3	4	5	6	7
Coordination									
D1	The collaboration was explicitly verbalized or discussed.	n/a	1	2	3	4	5	6	7
D2	The collaboration was written down in detail.	n/a	1	2	3	4	5	6	7

		Strongly Disagree						Strongly Agree
D3	Deliverables were clearly defined.	n/a	1	2	3	4	5	6	7
D4	Our organization's programs were well-coordinated with the network's programs.	n/a	1	2	3	4	5	6	7
D5	Our activities with this network were well coordinated.	n/a	1	2	3	4	5	6	7
D6	We felt like we never knew what we are supposed to be doing for the collaboration.	n/a	1	2	3	4	5	6	7
D7	We felt like we never knew when we were supposed to be contributing to the collaboration.	n/a	1	2	3	4	5	6	7
D8	There was an individual, group or organization that took responsibility for the collaboration who was expected to take care of coordinating activities in the network and also exercising authority on behalf of the network if necessary.	n/a	1	2	3	4	5	6	7
D9	A coordinating body was designated or identified that includes input from all collaborators.	n/a	1	2	3	4	5	6	7
D10	A coordinating body ensured that all collaborators were working in synchronization.	n/a	1	2	3	4	5	6	7
D11	The role of the 'network manager' was of more of a coordinator than of traditional management characterized by hierarchies, bureaucracy, centralization and opportunism.	n/a	1	2	3	4	5	6	7

Communication Efficiency

E1	The other participants were unable to transmit information that was required through the network.	n/a	1	2	3	4	5	6	7
E2	The other participants were unwilling to transmit information that was expected through the network.	n/a	1	2	3	4	5	6	7
E3	Information was not transmitted through the network because it was not valuable enough to be transmitted.	n/a	1	2	3	4	5	6	7
E4	Information that we received via the collaboration lead to a change in knowledge.	n/a	1	2	3	4	5	6	7
E5	Information that we received via the collaboration lead to a change in attitude.	n/a	1	2	3	4	5	6	7
E6	Information that we received via the collaboration lead to a change in behaviour.	n/a	1	2	3	4	5	6	7
E7	There were problems identifying the relevant persons to transfer information to or to obtain information from.	n/a	1	2	3	4	5	6	7
E8	There was an understanding of the inputs made and progress of the collaboration.	n/a	1	2	3	4	5	6	7
E9	Communication in the network was transparent.	n/a	1	2	3	4	5	6	7
E10	Communication in the network was clear and accessible.	n/a	1	2	3	4	5	6	7
E11	There were problems of knowledge credibility in the collaboration.	n/a	1	2	3	4	5	6	7
E12	There were shared understanding of the meaning of information transferred among participants.	n/a	1	2	3	4	5	6	7
E13	Information communicated by participants was not used.	n/a	1	2	3	4	5	6	7
E14	Communication in the network was too costly.	n/a	1	2	3	4	5	6	7
E15	There were no secrecy problems in the network.	n/a	1	2	3	4	5	6	7
E16	There were no secrecy breaches in the network.	n/a	1	2	3	4	5	6	7
E17	There were information leaks in the collaborative network.	n/a	1	2	3	4	5	6	7
E18	Difficulty was experienced in getting ideas clearly across to other collaborators when communication was made with them.	n/a	1	2	3	4	5	6	7
E19	When there was a need to communicate with other collaborators, there was difficulty in contacting them.	n/a	1	2	3	4	5	6	7

		Strongly Disagree							Strongly Agree
R&D Efficiency									
F1	The collaboration in the network was productive.	n/a	1	2	3	4	5	6	7
F2	The collaboration generated sufficient outputs for the investments of resources.	n/a	1	2	3	4	5	6	7
F3	The collaboration resulted in value for money.	n/a	1	2	3	4	5	6	7
F4	The time spent in the collaboration was worthwhile.	n/a	1	2	3	4	5	6	7
F5	We were always delighted with the performance coming out of this network.	n/a	1	2	3	4	5	6	7
F6	The outcomes from the collaboration justified expenditures.	n/a	1	2	3	4	5	6	7
F7	Compensation of participants was linked to their deliverables.	n/a	1	2	3	4	5	6	7
F8	Participants in the collaboration obtained fair returns based on the value that they contributed.	n/a	1	2	3	4	5	6	7

Network Effectiveness

The following question relates to the level of network effectiveness that you perceive.

Using the following scale, in your view please indicate the level of effectiveness of this network. Please **circle one percentage** (%) figure. Zero percent (0%) indicates the lowest level of effectiveness and one hundred percent (100%) indicates the highest level of effectiveness.

0%   10%   20%   30%   40%   50%   60%   70%   80%   90%   100%

Partner Identification

Please place and X next to the organizations from the list with which your organization had direct relationships over a continuous basis rather than one-off projects. You may also include the names of organizations in the provided space which are not included in the list.

☐ Australian Wine Research Institute (AWRI)

☐ Grape and Wine Research Development Corporation (GWRDC)

☐ Commonwealth Scientific and Industrial Research Organization (CSIRO)

☐ South Australian Research and Development Institute (SARDI)

☐ Cooperative Research Centres

☐ Universities

☐ Wine Companies

☐ Other

Relationship Characteristics

Please complete the following form, one for **each organization** with which you had **direct** relationships that you identified in the previous section.

Name of Organization ....

Organization Type ☐ Business ☐ Government ☐ University ☐ Research Organization

Name of Contact Person ....

Length of relationship

			Strongly Disagree						Strongly Agree
Trust									
G1	This partner kept promises it made to our organization.	n/a	1	2	3	4	5	6	7
G2	This partner was not always honest with us.	n/a	1	2	3	4	5	6	7
G3	We believed the information that this partner provided us.	n/a	1	2	3	4	5	6	7
G4	The partner was genuinely concerned that our efforts succeeded.	n/a	1	2	3	4	5	6	7
G5	When making important decisions, this partner considered our welfare as well as its own.	n/a	1	2	3	4	5	6	7
G6	We trusted this partner to keep our best interests in mind.	n/a	1	2	3	4	5	6	7
G7	This partner was trustworthy.	n/a	1	2	3	4	5	6	7
G8	We found it necessary to be cautious with this partner.	n/a	1	2	3	4	5	6	7
G9	This partner made sacrifices for us in the past.	n/a	1	2	3	4	5	6	7
G10	We felt that this partner was on our side.	n/a	1	2	3	4	5	6	7
G12	This partner was frank in dealing with us.	n/a	1	2	3	4	5	6	7
G13	This partner could be counted on to do what is right.	n/a	1	2	3	4	5	6	7
G14	In our relationship, this partner had high integrity.	n/a	1	2	3	4	5	6	7
Commitment									
H1	We defended this partner when others criticized.	n/a	1	2	3	4	5	6	7
H2	We had a strong sense of loyalty to this partner.	n/a	1	2	3	4	5	6	7
H3	We were continually on the lookout for another partner to replace this one.	n/a	1	2	3	4	5	6	7
H4	We expected to work with this partner for some time.	n/a	1	2	3	4	5	6	7
H5	If another partner offered us better service, we would most certainly take them on, even if it meant dropping this partner.	n/a	1	2	3	4	5	6	7
H6	We were not very committed to this partner.	n/a	1	2	3	4	5	6	7
H7	We were quite willing to make long-term investment in our relationship to this partner.	n/a	1	2	3	4	5	6	7
H8	The relationship with this partner deserved our effort to maintain it.	n/a	1	2	3	4	5	6	7
H9	We were patient with this partner when they made mistakes that caused us trouble.	n/a	1	2	3	4	5	6	7

If you have any further comment, please include it in the following space:

Relationship Characteristics

Please complete the following form, one for **each organization** with which you had **direct** relationships that you identified in the previous section.

Name of Organization

Organization Type ☐ Business ☐ Government ☐ University ☐ Research Organization

Name of Contact Person

Length of relationship

			Strongly Disagree						Strongly Agree
Trust									
G1	This partner kept promises it made to our organization.	n/a	1	2	3	4	5	6	7
G2	This partner was not always honest with us.	n/a	1	2	3	4	5	6	7
G3	We believed the information that this partner provided us.	n/a	1	2	3	4	5	6	7
G4	The partner was genuinely concerned that our efforts succeeded.	n/a	1	2	3	4	5	6	7
G5	When making important decisions, this partner considered our welfare as well as its own.	n/a	1	2	3	4	5	6	7
G6	We trusted this partner to keep our best interests in mind.	n/a	1	2	3	4	5	6	7
G7	This partner was trustworthy.	n/a	1	2	3	4	5	6	7
G8	We found it necessary to be cautious with this partner.	n/a	1	2	3	4	5	6	7
G9	This partner made sacrifices for us in the past.	n/a	1	2	3	4	5	6	7
G10	We felt that this partner was on our side.	n/a	1	2	3	4	5	6	7
G12	This partner was frank in dealing with us.	n/a	1	2	3	4	5	6	7
G13	This partner could be counted on to do what is right.	n/a	1	2	3	4	5	6	7
G14	In our relationship, this partner had high integrity.	n/a	1	2	3	4	5	6	7
Commitment									
H1	We defended this partner when others criticized.	n/a	1	2	3	4	5	6	7
H2	We had a strong sense of loyalty to this partner.	n/a	1	2	3	4	5	6	7
H3	We were continually on the lookout for another partner to replace this one.	n/a	1	2	3	4	5	6	7
H4	We expected to work with this partner for some time.	n/a	1	2	3	4	5	6	7
H5	If another partner offered us better service, we would most certainly take them on, even if it meant dropping this partner.	n/a	1	2	3	4	5	6	7
H6	We were not very committed to this partner.	n/a	1	2	3	4	5	6	7
H7	We were quite willing to make long-term investment in our relationship to this partner.	n/a	1	2	3	4	5	6	7
H8	The relationship with this partner deserved our effort to maintain it.	n/a	1	2	3	4	5	6	7
H9	We were patient with this partner when they made mistakes that caused us trouble.	n/a	1	2	3	4	5	6	7

If you have any further comment, please include it in the following space:

Relationship Characteristics

Please complete the following form, one for **each organization** with which you had **direct** relationships that you identified in the previous section.

Name of Organization _____

Organization Type ☐ Business ☐ Government ☐ University ☐ Research Organization

Name of Contact Person _____

Length of relationship

			Strongly Disagree					Strongly Agree	
Trust									
G1	This partner kept promises it made to our organization.	n/a	1	2	3	4	5	6	7
G2	This partner was not always honest with us.	n/a	1	2	3	4	5	6	7
G3	We believed the information that this partner provided us.	n/a	1	2	3	4	5	6	7
G4	The partner was genuinely concerned that our efforts succeeded.	n/a	1	2	3	4	5	6	7
G5	When making important decisions, this partner considered our welfare as well as its own.	n/a	1	2	3	4	5	6	7
G6	We trusted this partner to keep our best interests in mind.	n/a	1	2	3	4	5	6	7
G7	This partner was trustworthy.	n/a	1	2	3	4	5	6	7
G8	We found it necessary to be cautious with this partner.	n/a	1	2	3	4	5	6	7
G9	This partner made sacrifices for us in the past.	n/a	1	2	3	4	5	6	7
G10	We felt that this partner was on our side.	n/a	1	2	3	4	5	6	7
G12	This partner was frank in dealing with us.	n/a	1	2	3	4	5	6	7
G13	This partner could be counted on to do what is right.	n/a	1	2	3	4	5	6	7
G14	In our relationship, this partner had high integrity.	n/a	1	2	3	4	5	6	7
Commitment									
H1	We defended this partner when others criticized.	n/a	1	2	3	4	5	6	7
H2	We had a strong sense of loyalty to this partner.	n/a	1	2	3	4	5	6	7
H3	We were continually on the lookout for another partner to replace this one.	n/a	1	2	3	4	5	6	7
H4	We expected to work with this partner for some time.	n/a	1	2	3	4	5	6	7
H5	If another partner offered us better service, we would most certainly take them on, even if it meant dropping this partner.	n/a	1	2	3	4	5	6	7
H6	We were not very committed to this partner.	n/a	1	2	3	4	5	6	7
H7	We were quite willing to make long-term investment in our relationship to this partner.	n/a	1	2	3	4	5	6	7
H8	The relationship with this partner deserved our effort to maintain it.	n/a	1	2	3	4	5	6	7
H9	We were patient with this partner when they made mistakes that caused us trouble.	n/a	1	2	3	4	5	6	7

If you have any further comment, please include it in the following space:

Relationship Characteristics

Please complete the following form, one for **each organization** with which you had **direct** relationships that you identified in the previous section.

Name of Organization

Organization Type ☐ Business ☐ Government ☐ University ☐ Research Organization

Name of Contact Person

Length of relationship

			Strongly Disagree						Strongly Agree
Trust									
G1	This partner kept promises it made to our organization.	n/a	1	2	3	4	5	6	7
G2	This partner was not always honest with us.	n/a	1	2	3	4	5	6	7
G3	We believed the information that this partner provided us.	n/a	1	2	3	4	5	6	7
G4	The partner was genuinely concerned that our efforts succeeded.	n/a	1	2	3	4	5	6	7
G5	When making important decisions, this partner considered our welfare as well as its own.	n/a	1	2	3	4	5	6	7
G6	We trusted this partner to keep our best interests in mind.	n/a	1	2	3	4	5	6	7
G7	This partner was trustworthy.	n/a	1	2	3	4	5	6	7
G8	We found it necessary to be cautious with this partner.	n/a	1	2	3	4	5	6	7
G9	This partner made sacrifices for us in the past.	n/a	1	2	3	4	5	6	7
G10	We felt that this partner was on our side.	n/a	1	2	3	4	5	6	7
G12	This partner was frank in dealing with us.	n/a	1	2	3	4	5	6	7
G13	This partner could be counted on to do what is right.	n/a	1	2	3	4	5	6	7
G14	In our relationship, this partner had high integrity.	n/a	1	2	3	4	5	6	7
Commitment									
H1	We defended this partner when others criticized.	n/a	1	2	3	4	5	6	7
H2	We had a strong sense of loyalty to this partner.	n/a	1	2	3	4	5	6	7
H3	We were continually on the lookout for another partner to replace this one.	n/a	1	2	3	4	5	6	7
H4	We expected to work with this partner for some time.	n/a	1	2	3	4	5	6	7
H5	If another partner offered us better service, we would most certainly take them on, even if it meant dropping this partner.	n/a	1	2	3	4	5	6	7
H6	We were not very committed to this partner.	n/a	1	2	3	4	5	6	7
H7	We were quite willing to make long-term investment in our relationship to this partner.	n/a	1	2	3	4	5	6	7
H8	The relationship with this partner deserved our effort to maintain it.	n/a	1	2	3	4	5	6	7
H9	We were patient with this partner when they made mistakes that caused us trouble.	n/a	1	2	3	4	5	6	7

If you have any further comment, please include it in the following space:

Appendix D: Online Questionnaire for Final Field Work

Note: For readability, only a screenshot of the first page will be included. The other pages of the questionnaire will be displayed in Word format rather than the online display. This is also justified given the inability to view the entire webpage at once as some of the content could only be viewed by scrolling.

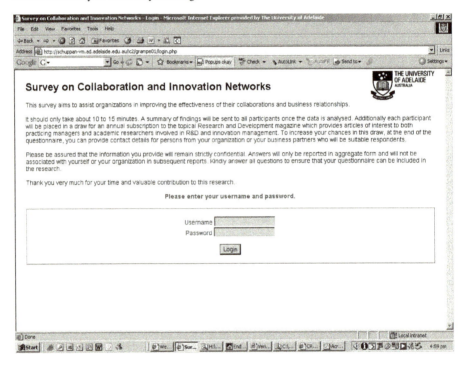

Survey Questionnaire on Innovation Networks

Details

Your Details

Please fill in the following details.

Name _____

Organization _____

Please place an X in **one or more** of the squares to indicate your answer to each question.

Organization Type: ☐ Business ☐ Government ☐ University ☐ Research Organization

Industry Sector: ☐ Biotechnology ☐ ICT ☐ Wine ☐ Other _____

Location of your office:

☐ Australian Capital Territory ☐ Queensland ☐ Victoria

☐ New South Wales ☐ Tasmania ☐ Northern Territory

☐ South Australia

Clarification of Terms and References

Network: Group of organizations that continuously interact with each other comprising relationships that vary in formality from contracting to informal relationships. The terms network, collaboration and collaborative network are used interchangeably.

Participant: An organization belonging to the network. It may be a government agency, research organization, university or business.

Example: The following diagram may reflect an example of a network in which your organization may have been a participant. You may be involved in several different networks. Questions on the next page relate to ONE innovation network in which you in the capacity of your role on behalf of your organization were involved.

Network Processes

The following statements relate to processes of the network. Please place an X next to **one response per statement.**

		Strongly Disagree							Strongly Agree
A1	One or more large participants dominated the network.	☐ n/a ☐ 1	☐ 2	☐ 3	☐ 4	☐ 5	☐ 6	☐ 7	
A2	The power distribution in the network was even.	☐ n/a ☐ 1	☐ 2	☐ 3	☐ 4	☐ 5	☐ 6	☐ 7	
A3	My organization had the same amount of power as the other participants' organizations.	☐ n/a ☐ 1	☐ 2	☐ 3	☐ 4	☐ 5	☐ 6	☐ 7	
A4	A particular participant had tremendous influence over the other players in the network.	☐ n/a ☐ 1	☐ 2	☐ 3	☐ 4	☐ 5	☐ 6	☐ 7	
A5	A particular participant had more control than others over resources such as funding, equipment, skills and competencies.	☐ n/a ☐ 1	☐ 2	☐ 3	☐ 4	☐ 5	☐ 6	☐ 7	
B1	During negotiation, meetings or discussions, there was give-and-take among participants. Each challenged the others and tried to understand the others points of view.	☐ n/a ☐ 1	☐ 2	☐ 3	☐ 4	☐ 5	☐ 6	☐ 7	
B2	The research institution and the industry partner were involved in the early phases of discussion in setting the research agenda.	☐ n/a ☐ 1	☐ 2	☐ 3	☐ 4	☐ 5	☐ 6	☐ 7	
B3	Conflicts between participants were resolved locally among the disagreeing participants rather than via escalation throughout the wider network.	☐ n/a ☐ 1	☐ 2	☐ 3	☐ 4	☐ 5	☐ 6	☐ 7	
B4	There was compromise among participants in decision-making and each party obtained value from the network.	☐ n/a ☐ 1	☐ 2	☐ 3	☐ 4	☐ 5	☐ 6	☐ 7	
C1	The collaboration was explicitly verbalized or discussed.	☐ n/a ☐ 1	☐ 2	☐ 3	☐ 4	☐ 5	☐ 6	☐ 7	
C2	The collaboration was written down in detail.	☐ n/a ☐ 1	☐ 2	☐ 3	☐ 4	☐ 5	☐ 6	☐ 7	
D3	Our organization's programs were well-coordinated with the network's programs.	☐ n/a ☐ 1	☐ 2	☐ 3	☐ 4	☐ 5	☐ 6	☐ 7	
C4	Our activities with this network were well coordinated.	☐ n/a ☐ 1	☐ 2	☐ 3	☐ 4	☐ 5	☐ 6	☐ 7	
C5	We felt like we never knew what we are supposed to be doing for the collaboration.	☐ n/a ☐ 1	☐ 2	☐ 3	☐ 4	☐ 5	☐ 6	☐ 7	
C6	We felt like we never knew when we were supposed to be contributing to the collaboration.	☐ n/a ☐ 1	☐ 2	☐ 3	☐ 4	☐ 5	☐ 6	☐ 7	
C7	There was an individual, group or organization *(either existing or new)* that took responsibility for the collaboration who was expected to take care of coordinating activities in the network and also exercising authority on behalf of the network if necessary.	☐ n/a ☐ 1	☐ 2	☐ 3	☐ 4	☐ 5	☐ 6	☐ 7	
C8	A coordinating body was designated or identified that includes input from all collaborators.	☐ n/a ☐ 1	☐ 2	☐ 3	☐ 4	☐ 5	☐ 6	☐ 7	
C9	A coordinating body ensured that all collaborators were working in synchronization.	☐ n/a ☐ 1	☐ 2	☐ 3	☐ 4	☐ 5	☐ 6	☐ 7	
C10	The role of the 'network manager' was of more of a coordinator than of traditional management characterized by hierarchies, bureaucracy, centralization and opportunism.	☐ n/a ☐ 1	☐ 2	☐ 3	☐ 4	☐ 5	☐ 6	☐ 7	

D1	The other participants were unable to transmit information that was required through the network.	☐ n/a ☐ 1 ☐ 2 ☐ 3 ☐ 4 ☐ 5 ☐ 6 ☐ 7
D2	The other participants were unwilling to transmit information that was expected through the network.	☐ n/a ☐ 1 ☐ 2 ☐ 3 ☐ 4 ☐ 5 ☐ 6 ☐ 7
D3	Information that we received via the collaboration lead to a change in knowledge.	☐ n/a ☐ 1 ☐ 2 ☐ 3 ☐ 4 ☐ 5 ☐ 6 ☐ 7
D4	Information that we received via the collaboration lead to a change in attitude.	☐ n/a ☐ 1 ☐ 2 ☐ 3 ☐ 4 ☐ 5 ☐ 6 ☐ 7
D5	Information that we received via the collaboration lead to a change in behaviour.	☐ n/a ☐ 1 ☐ 2 ☐ 3 ☐ 4 ☐ 5 ☐ 6 ☐ 7
D6	There were problems identifying the relevant persons to transfer information to or to obtain information from.	☐ n/a ☐ 1 ☐ 2 ☐ 3 ☐ 4 ☐ 5 ☐ 6 ☐ 7
D7	There was an understanding of the inputs made and progress of the collaboration.	☐ n/a ☐ 1 ☐ 2 ☐ 3 ☐ 4 ☐ 5 ☐ 6 ☐ 7
D8	Communication in the network was transparent.	☐ n/a ☐ 1 ☐ 2 ☐ 3 ☐ 4 ☐ 5 ☐ 6 ☐ 7
D9	Communication in the network was clear and accessible.	☐ n/a ☐ 1 ☐ 2 ☐ 3 ☐ 4 ☐ 5 ☐ 6 ☐ 7
D10	There were problems of knowledge credibility in the collaboration.	☐ n/a ☐ 1 ☐ 2 ☐ 3 ☐ 4 ☐ 5 ☐ 6 ☐ 7
D11	There were shared understanding of the meaning of information transferred among participants.	☐ n/a ☐ 1 ☐ 2 ☐ 3 ☐ 4 ☐ 5 ☐ 6 ☐ 7
D12	Information communicated by participants was not used.	☐ n/a ☐ 1 ☐ 2 ☐ 3 ☐ 4 ☐ 5 ☐ 6 ☐ 7
D13	Communication in the network was too costly.	☐ n/a ☐ 1 ☐ 2 ☐ 3 ☐ 4 ☐ 5 ☐ 6 ☐ 7
D14	There were no secrecy problems in the network.	☐ n/a ☐ 1 ☐ 2 ☐ 3 ☐ 4 ☐ 5 ☐ 6 ☐ 7
D15	There were no secrecy breaches in the network.	☐ n/a ☐ 1 ☐ 2 ☐ 3 ☐ 4 ☐ 5 ☐ 6 ☐ 7
D16	There were information leaks in the collaborative network.	☐ n/a ☐ 1 ☐ 2 ☐ 3 ☐ 4 ☐ 5 ☐ 6 ☐ 7
D17	Difficulty was experienced in getting ideas clearly across to other collaborators when communication was made with them.	☐ n/a ☐ 1 ☐ 2 ☐ 3 ☐ 4 ☐ 5 ☐ 6 ☐ 7
D18	When there was a need to communicate with other collaborators, there was difficulty in contacting them.	☐ n/a ☐ 1 ☐ 2 ☐ 3 ☐ 4 ☐ 5 ☐ 6 ☐ 7
E1	The collaboration in the network was productive.	☐ n/a ☐ 1 ☐ 2 ☐ 3 ☐ 4 ☐ 5 ☐ 6 ☐ 7
E2	The collaboration generated sufficient outputs for the investments of resources.	☐ n/a ☐ 1 ☐ 2 ☐ 3 ☐ 4 ☐ 5 ☐ 6 ☐ 7
E3	The collaboration resulted in value for money.	☐ n/a ☐ 1 ☐ 2 ☐ 3 ☐ 4 ☐ 5 ☐ 6 ☐ 7
E4	The time spent in the collaboration was worthwhile.	☐ n/a ☐ 1 ☐ 2 ☐ 3 ☐ 4 ☐ 5 ☐ 6 ☐ 7
E5	We were always delighted with the performance coming out of this network.	☐ n/a ☐ 1 ☐ 2 ☐ 3 ☐ 4 ☐ 5 ☐ 6 ☐ 7
E6	The outcomes from the collaboration justified expenditures.	☐ n/a ☐ 1 ☐ 2 ☐ 3 ☐ 4 ☐ 5 ☐ 6 ☐ 7

Network Effectiveness

The following question relates to the level of network effectiveness that you perceive. Using the following scale, in your view please indicate the level of effectiveness of this network. Please place an X next to **one percentage** (%) figure. Zero percent (0%) indicates the lowest level of effectiveness and one hundred percent (100%) indicates the highest level of effectiveness.

☐ 0% ☐ 10% ☐ 20% ☐ 30% ☐ 40% ☐ 50% ☐ 60% ☐ 70% ☐ 80% ☐ 90% ☐ 100%

161

Relationship Characteristics

Please complete the following form, one for **each organization** with which you had **direct** relationships that you identified in the previous section.

Name of Organization _____

Organization Type ☐ Business   ☐ Government   ☐ University   ☐ Research Organization

Name of Contact Person _____

Length of relationship

		Strongly Disagree						Strongly Agree
F1	This partner kept promises it made to our organization.	☐ n/a ☐ 1	☐ 2	☐ 3	☐ 4	☐ 5	☐ 6	☐ 7
F2	This partner was not always honest with us.	☐ n/a ☐ 1	☐ 2	☐ 3	☐ 4	☐ 5	☐ 6	☐ 7
F3	We believed the information that this partner provided us.	☐ n/a ☐ 1	☐ 2	☐ 3	☐ 4	☐ 5	☐ 6	☐ 7
F4	The partner was genuinely concerned that our efforts succeeded.	☐ n/a ☐ 1	☐ 2	☐ 3	☐ 4	☐ 5	☐ 6	☐ 7
F5	When making important decisions, this partner considered our welfare as well as its own.	☐ n/a ☐ 1	☐ 2	☐ 3	☐ 4	☐ 5	☐ 6	☐ 7
F6	We trusted this partner to keep our best interests in mind.	☐ n/a ☐ 1	☐ 2	☐ 3	☐ 4	☐ 5	☐ 6	☐ 7
F7	This partner was trustworthy.	☐ n/a ☐ 1	☐ 2	☐ 3	☐ 4	☐ 5	☐ 6	☐ 7
F8	We found it necessary to be cautious with this partner.	☐ n/a ☐ 1	☐ 2	☐ 3	☐ 4	☐ 5	☐ 6	☐ 7
F9	This partner made sacrifices for us in the past.	☐ n/a ☐ 1	☐ 2	☐ 3	☐ 4	☐ 5	☐ 6	☐ 7
F10	We felt that this partner was on our side.	☐ n/a ☐ 1	☐ 2	☐ 3	☐ 4	☐ 5	☐ 6	☐ 7
F12	This partner was frank in dealing with us.	☐ n/a ☐ 1	☐ 2	☐ 3	☐ 4	☐ 5	☐ 6	☐ 7
F13	This partner could be counted on to do what is right.	☐ n/a ☐ 1	☐ 2	☐ 3	☐ 4	☐ 5	☐ 6	☐ 7
F14	In our relationship, this partner had high integrity.	☐ n/a ☐ 1	☐ 2	☐ 3	☐ 4	☐ 5	☐ 6	☐ 7
G1	We defended this partner when others criticized.	☐ n/a ☐ 1	☐ 2	☐ 3	☐ 4	☐ 5	☐ 6	☐ 7
G2	We had a strong sense of loyalty to this partner.	☐ n/a ☐ 1	☐ 2	☐ 3	☐ 4	☐ 5	☐ 6	☐ 7
G3	We were continually on the lookout for another partner to replace this one.	☐ n/a ☐ 1	☐ 2	☐ 3	☐ 4	☐ 5	☐ 6	☐ 7
G4	We expected to work with this partner for some time.	☐ n/a ☐ 1	☐ 2	☐ 3	☐ 4	☐ 5	☐ 6	☐ 7
G5	If another partner offered us better service, we would most certainly take them on, even if it meant dropping this partner.	☐ n/a ☐ 1	☐ 2	☐ 3	☐ 4	☐ 5	☐ 6	☐ 7
G6	We were not very committed to this partner.	☐ n/a ☐ 1	☐ 2	☐ 3	☐ 4	☐ 5	☐ 6	☐ 7
G7	We were quite willing to make long-term investment in our relationship to this partner.	☐ n/a ☐ 1	☐ 2	☐ 3	☐ 4	☐ 5	☐ 6	☐ 7
G8	The relationship with this partner deserved our effort to maintain it.	☐ n/a ☐ 1	☐ 2	☐ 3	☐ 4	☐ 5	☐ 6	☐ 7
G9	We were patient with this partner when they made mistakes that caused us trouble.	☐ n/a ☐ 1	☐ 2	☐ 3	☐ 4	☐ 5	☐ 6	☐ 7

Appendix E: Assessment of Normality

Appendix E.1: Assessment of Normality for the ICT Industry

Variable	skewness	c.r.	kurtosis	c.r.
Trust	-.902	-3.590	.444	.884
Power	-.300	-1.194	-.638	-1.270
Harmony	-.875	-3.484	1.068	2.125
Coordination	-.645	-2.567	1.397	2.780
Communication efficiency	-.540	-2.150	.633	1.299
R&D efficiency	-.950	-3.780	.537	1.068
Effectiveness	-1.055	-4.196	.301	.599
Multivariate			6.182	2.684

Appendix E.2: Assessment of Normality for the B/N Industry

Variable	skewness	c.r.	kurtosis	c.r.
Trust	-.789	-3.585	.005	.012
Power	-.357	-1.622	-.457	-1.039
Harmony	.690	-3.139	-.308	-.700
Coordination	-.537	-2.443	-.057	-.129
Communication efficiency	.520	-2.365	.047	.106
R&D efficiency	-.598	-2.718	.059	.133
Effectiveness	-1.075	-4.888	.684	1.554
Multivariate			5.659	2.807

Appendix F: Congeneric Models

Appendix F.1: Congeneric Model – Trust – B/N network

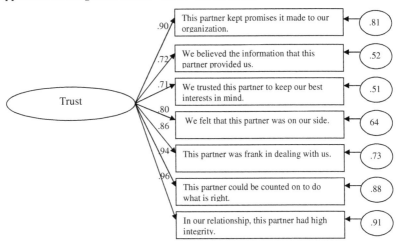

Appendix F.2: Congeneric Model – Trust – ICT network

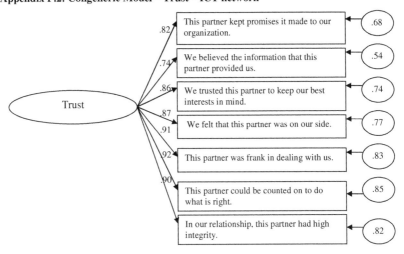

Appendix F.3: Congeneric Model – Power Distribution – B/N network

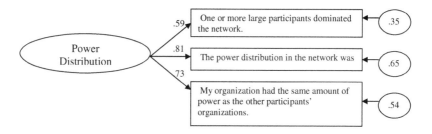

Appendix F.4: Congeneric Model – Power Distribution – ICT network

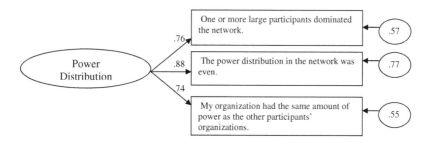

Appendix F.5: Congeneric Model – Coordination – B/N network

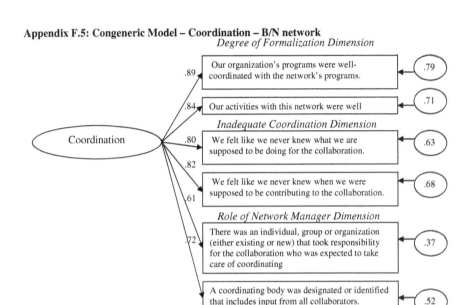

Appendix F.6: Congeneric Model – Coordination – ICT network

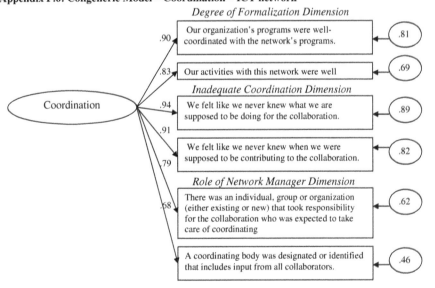

Appendix F.7: Congeneric Model – Harmony – B/N network

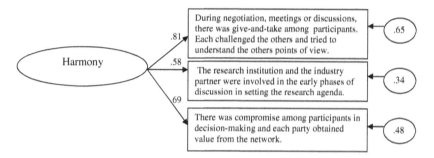

Appendix F.8: Congeneric Model – Harmony – ICT network

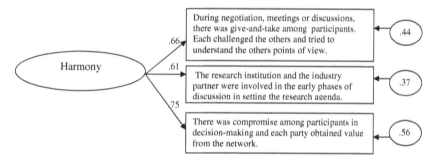

Appendix F.9: Congeneric Model – Communication Efficiency – B/N network

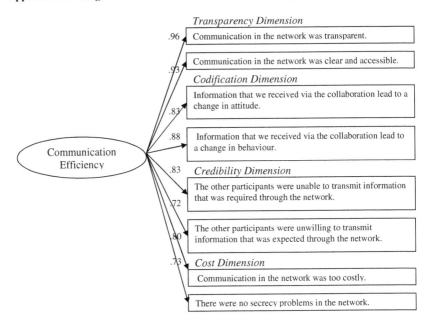

Transparency Dimension

.96 — Communication in the network was transparent.

Communication in the network was clear and accessible.

.93

Codification Dimension

Information that we received via the collaboration lead to a change in attitude.

.83

.88 — Information that we received via the collaboration lead to a change in behaviour.

.83 *Credibility Dimension*

The other participants were unable to transmit information that was required through the network.

.72

The other participants were unwilling to transmit information that was expected through the network.

.80

.73 *Cost Dimension*

Communication in the network was too costly.

There were no secrecy problems in the network.

Communication Efficiency

Appendix F.10: Congeneric Model – Communication Efficiency – ICT network

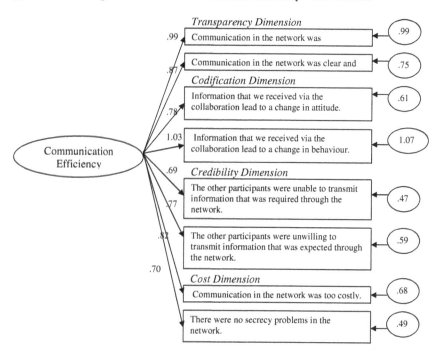

Appendix F.11: Congeneric Model – R&D Efficiency – B/N network

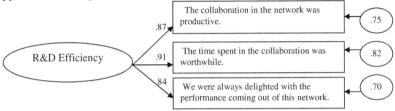

Appendix F.12: Congeneric Model – R&D Efficiency – ICT network

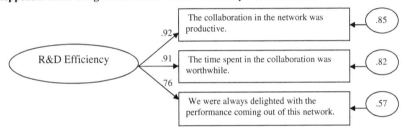

Appendix G: Correlation Matrices

Appendix G.1: Correlation Matrix - ICT Industry

	effectiveness	power	R&Deff	commEff	harmony	coordination	trust
effectiveness	1.000						
power	-.354	1.000					
R&D efficiency	.855	-.372	1.000				
commEff	.690	-.389	.725	1.000			
harmony	.620	-.428	.651	.682	1.000		
coordination	.536	-.444	.563	.590	.649	1.000	
trust	.576	-.332	.605	.634	.697	.434	1.000

Appendix G.1: Correlation Matrix - B/N Industry

	effectiveness	power	R&Deff	commEff	harmony	coordination	trust
effectiveness	1.000						
power	-.458	1.000					
R&Deff	.821	-.452	1.000				
commEff	.755	-.448	.676	1.000			
harmony	.736	-.481	.726	.719	1.000		
coordination	.646	-.469	.637	.632	.678	1.000	
trust	.675	-.464	.666	.660	.708	.535	1.000